The clerk and the accounts department

The clerk and the accounts department

The story of 6 mathematicians

Jack Dikian

First published 2019

Copyright © 2019 by Jack Dikian

Sydney, Summer 2019

For my mother and late father

"Sometimes I go about in pity for myself, and all the while, a great wind carries me across the sky."

— Ojibwe saying

AUTHOR'S NOTE

Kurt Friedrich Gödel, Alan Mathison Turing, Srinivasa Ramanujan, Grigori Yakovlevich Perelman, Andrew John Wiles and John Forbes Nash; Six men of the 20th century whose stories are of the great romantic tales of mathematics. An account of triumph and tragedy, of men of genius, baring intellectual brilliance and tenacity. They prevailed against incredible adversity and whose lives, in the case of two cut short at the height of their powers.

These men proved Fermat's Last Theorem, solved the Poincaré conjecture whilst shunning acclaim and prize. Discovered the extraordinary connection, the partition function, Pi and nontrivial patterns in modular arithmetic. One developed a machine that helped break the German Enigma, another invented Game theory, making economics a practical endeavour. Gödel proved that the mathematical methods in place since the time of Euclid inadequate for discovering all that is true about the natural numbers.

Contents

1. The lion and the mouse   8
2. The dog and the shadow   18
3. The zonda and the sun   51
4. The grasshopper and the bull   87
5. The wolf and the crane   114
6. The snowbird and the pitcher   148

1: The Lion and the Mouse

A Lion is fast asleep until he's abruptly woken by a pesty Mouse. He opens his jaws ready to swallow it, but the Mouse begs him to think it over: I may be useful to you in the future, says the Mouse. The Lion laughs at the idea but, nonetheless, lets it go. Sometime later, the Lion is caught in a trap by some Hunters. At that moment the Mouse walks by and notices the Lion. He walks up and chews the rope, casually, freeing the Lion with little bother.

A young Cambridge mathematician, shambolic with an awkward halting speech builds a code-breaking engine decades ahead of its time. A man predestined to a gentler, more tolerant and enlightened time. But as is so often the case his life ended much too early and it ended after much humiliation, suffering and pain. He was just 27-years-old. He was a genus. He was gay.

Before the lights go out, looking back in the house, you must have seen old faces, child-like with expectancy and delight. The door that keeps the unknown out lets it come in. The strangest things can happen now. Late at night, one of us sometimes has noticed, eyes glued, glazed, and the yellow hues of film a distant memory of some. Of the vivid figures quick upon the screen, surely by now all of them are dead.

The patient terrier sulks in the rain; no avail, benevolence hardened by existence. We will make our meek adjustments, awkward, contented with such random consolations. The wind deposits the fall and the yelping wire haired faithful no longer of one; inky dye night looming closer, the fright. Ignition far yonder carts wretchedness and toll.

A grail of laughter of an empty can reverberate mutely, sounds of gaiety and quest - forced smirks; signs of mood long departed. But now we have seen the doom, that inevitable thumb. War is upon us once more; pain of courage, a side-step to dominion but at what charge. A final leer makes no sense of all.

War was indeed nigh. At 11:15, September 3, 1939 Chamberlain, pensive, brooding, tells a nation, a world that his country is at war. To some few; the unemployed finally find work; war work, a steady income. No more handouts, no more soup kitchens. But elations like these last not. A 17-year-old teenager Second Class aboard the storm-tossed British battle cruiser Hood writes home. Dearest Mum, I can'nae eat and my heart's in my mouth. He would not eat again. The Hood sinks - all hands too, all hope as well.

On the far side of Europe, a young Red Army Private lines up for war. How is he to know a million would die? Mama, he writes, I have not known such fear. Mama, please pray to God. When I was at home I didn't believe in God, but now I think of him 40 times a day. I don't know where to hide my head. Papa and Mama, farewell, I will never see you again, farewell, farewell, farewell.

Eighty-five Million souls parish, the war ends, all those lives; did they ever exist. MacArthur cries to God. Build me a son, he writes. O Lord, who will be strong enough to know when he is weak, and brave enough to face himself when he is afraid, one who will be proud and unbending in honest defeat, and humble and gentle in victory.

The hugely successful 2014 film, The Imitation Game tells the story of British code breakers working at a newly establish Government Code and Cipher School at Bletchley Park. Their task was to break the Germans' nearly unbreakable cipher machine, the Enigma. The film doesn't track Alan Turing's life after 1952 so as a consequence we didn't get to see that on June 7, 1954, just 16 days before his 42nd birthday, the softly-spoken genius took his life by taking a bite of the forbidden fruit; an apple that he had dipped in cyanide.

We also didn't get a sense of Turing's spirituality, something that wasn't the Christian scheme of heaven and hell, rather an emerging partiality for a more rationale, a more orderly belief in doctrines accessible to reason and investigation - Rationale Buddhism. He off course knew well enough that as an accursed sodomite he was reviled by both the Bible-believing Christians and an antihomosexual Britain. That wasn't always the case however. A very young Turing would refer to Jesus Christ as "Our Lord" with reverence according to his mother and Alan's close friends Mr. & Mrs Newmann; Though, even then, he would remark that the higher power is incomputable and as a causality unable to be pinned down in a mathematical equation.

After the untimely death of his boyhood sweetheart Christopher, Turing set out to investigate the distinction between the material mechanism of the body and the mind. He becomes obsessed with unravelling the nature of consciousness, its structure and its origins. His correspondence with Mrs. Morcom, Christopher's mother, reveal, how he longed to understand what had become of Christopher, of that essential aspect of him; his mind. He would write in 1930 "When the body dies the mechanism of the body, holding the spirit is gone and the spirit finds a new body sooner or later, perhaps immediately." Here, Turing turns toward an aspect of Buddhism once again, He places trust on the belief actions of a person lead to a new existence after death.

In 1939, the German invasion looming, the Poles who had themselves worked on breaking the Enigma pass on their work to the British. In turn, the British establish the Government Code and Cipher School at Bletchley Park in Buckinghamshire. Young, brash and proper, the brightest mathematicians, linguists and engineers would gather there. The numbers rose as the war went on; from a relatively small team in 1938 to something like 10,000 people, code-breakers, the Women's Royal Naval Service, the Women's Auxiliary Air Force, posh debutantes working on the cross-indexing system. Only god knows what

Bletchley locals must have thought of the foray and more so by the eccentricities of mathematicians. It has been said, probably playfully if war wasn't afoot, that the secret establishment was in fact a lunatic asylum.

During the build-up to the fierce fighting at El Alamein in 1942, the Bletchley Park code breakers were listening. Rommel, it's said gave them the welcome news that his tanks had insufficient fuel to fight effectively. The code breakers themselves had played a leading role in bringing this situation about for weeks, their decrypts had been unmasking the cargoes of the German and Italian ships carrying Rommel's supplies across the Mediterranean. This enabled the Royal Air Force to pick and choose the best targets. Thousands of tons of fuel were lost at sea. Some days into a battle, a broken message from Rommel confesses that the gradual annihilation of the army must be faced. Hitler responds, "Don't yield a single step; take the road leading to death or victory."

Churchill, Britain's wartime leader, will later declare; "The only thing that ever really frightened me during the war was the U-boat peril and the code breaker's work had at last made it possible to defuse that." According to historian of British intelligence if the U-boat Enigma had not been broken, and the war had continued for another two to three years, a further 14 to 21 million people might have been killed.

Altering the Enigma settings would only take a few minutes. The Germans would set their machines by predetermined and distributed code books - unknown to everyone but the intended recipient. And here lied the Deus ex machine. The Enigma had 159,000,000,000,000,000,000 possible positions. The code breakers, on the other hand calculated that to try every possible setting would take an eternity. If they had 10 men checking one setting a minute for 24 hours every day, seven days every week it would take roughly 20 Million years. They would have had to check 20 Million years of settings in just 20 minutes. That's not all. The Germans changed the Enigma settings at midnight.

It was obvious that a brute-force, exhaustive search was impractical. But by now the 27-year-old; who having alienated his colleagues confronts his Commanding Officer, Denniston. He wants £100,000 to build a machine. His boss, Alexander has already rejected the idea. Denniston, having long lost patience with Turing's proclivities dismisses him, too, with little regard and much scorn. if Turing is anything, he is persistent, but more than that.

One of the most influential figures of the 20th century is also the least understood. His self-imposing routines, all absorbing narrow preoccupations help, in a way, to protect the man. Offense and un-pleasantries carry little sway. Turing again, "the Enigma is an extremely well-designed machine – our problem is that we're only using men who are trying to beat it. What if only a machine can defeat another?"

The fortunes of the battle swayed. Turing's cipher machines (Bombes) were one of the most complex electro-mechanical devices of the time. His Bombes turned Bletchley Park into a codebreaking factory. As early as 1943 Turing's Bombes were breaking a staggering total of 84,000 messages each month - two messages every minute. The faster the messages could be broken, the fresher the intelligence that they contained, and on at least one occasion an intercepted message was being read at the British Admiralty less than 15 minutes after the Germans had transmitted it.

Some have speculated that Turing's rejection of his Christian faith was triggered by Christopher's death. Whatever the reason, it must be said that the intersection of the Buddhist philosophy with Turing's life-long yearning to make sense of the mind and expressed as the computational Theory of Mind, is arguably as important as his achievements at Bletchley Park.

The distinction Turing makes between the discarded mechanism of the body and the mind sets Turing on a ground-breaking path; To formulate a simple, logically complete and philosophically coherent definition of mechanism. A definition that he can now quite literally implement on his 1936 machine (a mathematical model of computation that represents an abstract machine) known today as the Turing machine. Because of its simplicity, the Turing machine demonstrates the fundamental capabilities and limitations of all physical systems in a clearly recognisable form. Turing's machine enables philosophers to develop a clear demarcation between mechanism and mind.

Here now is an important connexion. The philosophical notions of "easy - hard" problems are deconstructed. The hard problem of consciousness is the problem of explaining how and why we have qualitative phenomenal experiences. This is contrasted with the easy problems of explaining the ability to discriminate, integrate information, report mental states, focus attention, etc. Easy problems are easy because all that is required for their solution is to specify a mechanism that can perform the function. Turing's machine, as we know, can process all mechanisms. More so, Buddhist teachings help make his lifelong search to locate the mind that much clearer. By distilling out and coding mechanisms, what he finds left are in essence Buddhist modes of existential dependence. After all, it is hard to see how the mere enaction of a computation should give rise to an inner subjective life.

The limitations of qualitative experience is important both in computability theory and fundamental, and indeed fatal to the more modern idea; computationalism and hence materialism. Separating mechanism from intentionality and subjective conscious experience now underpins modern computability theory, artificial intelligence, computational neuroscience and has helped reframe cultural and philosophical views such as modernity, loss of certainty and moral relativism.

From the point of view of modern computability theory, Turing's machine can prove what machines can't do. By implication, what all computers can't do. An example of this is the important halting problem. The problem of determining, from a description of an arbitrary computer program and an input, whether the program will ever finish running, or halting. The historic importance of the halting problem is self-explanatory since it was considered to be one of the very first problems which was proven to be unsolvable. Over the passage of time, multiple other problems have emerged with solutions that are not yet known – or decided.

Today, there are no algorithms that correctly determines whether arbitrary programs eventually halt when run. The halting problem is said to be undecidable in polynomial time or NP-complete. A major unsolved problem in theoretical computer science is the P versus NP problem; that is If the solution to a problem can be verified in polynomial time, can it be found in polynomial time? The significance of this is such that it sits amongst another 6 problems laid out by the Clay Mathematics Institute in 2000. They're not easy – a correct solution to any one results in a US$1,000,000 prize being awarded by the institute.

Today, like Turing, many find Abrahamic religions to have failed them. They have found a faith such as Christianity for example unable to tackle, materialism, the tendency to consider material possessions and physical comfort as more important than spiritual values. instead they are degenerating into idiocy and bigotry. A wounded Turing sees in Buddhism an object of refuge for the intellectual. For those seeking spiritualty and for those wanting to escape the bleak and barren consequences of materialism. Turing's 1936 paper on the computable real numbers marks the epistemological break between idealism and materialism in mathematics. Prior to Turing it was hard to get away from the idea that through mathematical reason, the human mind gained access to a higher domain.

One advantage of establishing a rational basis for Buddhism is that it gives Buddhism an intellectual respectability at a time when the intellectual prestige of other religions is in steep decline, due to increasing obscurantism, which takes variety of forms varying from creationist anti-science to outright sabotage. When we dwell deeper into Turing's mind it's easy to see what one of the world's greatest mathematicians saw in Buddhism – an amalgam of spirituality accessible to reason and investigation in terms of shared, repeatable experience.

2: The dog and the shadow

A dog is walking home with a piece of meat in his mouth. On his way he crosses a river and looks into the water. He mistakes his own reflection for another dog that wants his dinner. But as he opens his mouth he lets it fall into the water.

Dear Sir,

I beg to introduce myself to you as a clerk in the Accounts Department of the Port Trust Office at Madras on a salary of only £20 per annum. I am now about 23 years of age. I have had no University education but I have undergone the ordinary school course. After leaving school I have been employing the spare time at my disposal to work at Mathematics. I have not trodden through the conventional regular course which is followed in a University course, but I am striking out a new path for myself. I have made a special investigation of divergent series in general and the results I get are termed by the local mathematicians as 'startling'.

I would request you to go through the enclosed papers. Being poor, if you are convinced that there is anything of value I would like to have my theorems published. I have not given the actual investigations nor the expressions that I get but I have indicated the lines on which I proceed. Being inexperienced I would very highly value any advice you give me. Requesting to be excused for the trouble I give you.

*Yours very sincerely
S. Ramanujan*

Srinivasa Ramanujan's story is one of the great romantic tales of mathematics. It is an account of triumph and tragedy, of a man of genius who prevailed against incredible adversity and whose life was cut short at the height of his powers. Ramanujan, a self-taught mathematician born in a rural village in South India, is said to have spent so much time thinking about mathematics that he failed out of college, the University of Madras twice. His wife and family would feed him almost by hand so that didn't have to stop doing his sums.

A deeply religious Hindu, Ramanujan credited his substantial mathematical capacities to divinity, and stated that the mathematical knowledge he displayed was revealed to him by his family goddess. "An equation for me has no meaning," he once said, "unless it expresses a thought of God." But Ramanujan's determination to lift himself from abject poverty, to be accepted, acknowledged as a first-class mathematician was not once at the cost of the social customs and caste prejudices of early 1900's.

From the time he travelled to Trinity College, Cambridge, to work with the mathematician and tutor Godfrey

Harold Hardy and John Edensor Littlewood on March 17, 1914 to 1919 when ill health compelled Ramanujan's return to India where he died in 1920 at the age of 32 he had achieved his lifelong ambition; to have his ideas, his maths by be validated by established mathematicians. Although he was extremely keen to have his work published as indicated by his early letters, he was also the first to acknowledge his limitations.

Some 50 years later, Janakiammal, his wife would say:

> "I considered it my good fortune to give him rice, lemon juice, butter milk, etc., at regular intervals and to give fomentation to his legs and chest when he reported pain. The two vessels used then for preparing hot water are alone still with me; these remind me often of those days."

During his short life, Ramanujan independently compiled nearly 4000 identities and equations, many completely novel, original, and highly unconventional.

Eventually, though, the frustrated Ramanujan spiralled into depression and illness, even attempting suicide at one time. His last letters to Hardy, written January 1920, show that he was still continuing to produce new ideas and theorems. His "lost notebook", really a collection of loose and unordered sheets of paper - more than one hundred pages written on 138 sides in Ramanujan's distinctive handwriting were rediscovered by George Andrews, the American mathematician, in 1976. It took almost 50 years to discover these papers because they were in a box of effects of G. N. Watson stored at the Wren Library at Trinity College, Cambridge.

Andrews' account of the discovery of the notebooks is in itself a story of perseverance, luck, good timing and intrigue. Andrews was due to attend a European conference in Strasbourg, when he obtained permission and support from the Trinity College library and from his professor, Ben Noble, to visit Cambridge after the conference, in order to investigate the unpublished writings of Watson et al. Noble agreed and with that commenced a remarkable tale.

Although not labelled as such, the identity of the papers was settled because Ramanujan's final letters to Hardy had referred to the discovery of what Ramanujan called mock theta functions, although without great detail, and the manuscript included what appeared to be his full notes on these. The majority of the formulas are about q-series and mock theta functions, about a third are about modular equations and singular moduli, and the remaining formulas are mainly about integrals, Dirichlet series, congruences, and asymptotics.

Hardy didn't just discover Ramanujan, his care in mentoring Ramanujan must be admired. Particularly knowing that Hardy was only a few years older than Ramanujan. Hardy's main worry was how to teach this astounding talent mathematics without destroying his confidence. The last thing Hardy wanted was to dent Ramanujan's fearless approach to the most difficult problems. Hardy would write:

> *"The limitations of his knowledge were as startling as its profundity. Here was a man who could work out modular equations, and*

theorems of complex multiplication, to orders unheard of, whose mastery of continued fractions was, on the formal side at any rate, beyond that of any mathematician in the world ... It was impossible to ask such a man to submit to systematic instruction, to try to learn mathematics from the beginning once more."

Hardy lived on for some 27 years after Ramanujan's death, to the ripe old age of 70. When asked in an interview what his greatest contribution to mathematics was, Hardy unhesitatingly replied that it was the discovery of Ramanujan, and even called their collaboration;

"the one romantic incident in my life."

And his death is one of the worst blows I have ever felt. But now I say to myself when I'm depressed and I find myself forced to listen to tiresome and pompous people, well, I have done something you have never done. I have collaborated with both Littlewood and Ramanujan's on something like equal terms. However, Hardy too became depressed later in life and attempted suicide by an overdose at one point. Some have blamed the Riemann Hypothesis for Ramanujan and Hardy's instabilities, giving it something of the reputation of a curse.

Madras 1913

Immediately before the War Madras was a thriving urban metropolis; a central administrative centre for the British in India's south. Flanked by a seaport township and linked by The Madras Railway Company to Mumba in the west and the tea plantations of Bengaluru in the South. Madras was one of only three major cities having a Telephone Exchange and was a city well on its way to becoming one of south India's largest cultural, economic and educational centres.

In 1913 Moses and Company, the tailors on Mount Road, were advertising their woollen suits and woollen underwear for Europe-bound students. Madras Corporation was debating the closure of a road in Santhome, others, many more likely preoccupied by the Golu or festive display of dolls in the late autumn year. Travelers, merchants, mostly British, mostly men of the empire would take tea at the old Spencer's hotel. Under lethargic, unhurried fans, birds chirping in the shade of the Fish Tail palms, affairs of the subcontinent would be conferred. Outside, women in their cool saris and children ribboned were on their way to school. Schools, at least according to finicky British-India records, were to be benefiting from a [Christian] missionary British education.

This fact has some bearing on how some students integrated their faith, undoubtedly Hindi, with Christian teachings, as well as how this formed their outlook on science and mathematics. Years later, one such student would wright "I discovered a wonderful synthesis between my faith and the cutting edge of science. It helped me to understand that science and spirituality aren't mutually exclusive. I also saw how much faith we place in the field of science and its theories."

Madras, old but otherwise in good condition according to The Times of India compered an unlikely union. Madras may well have been in good condition but that was really a matter of perspective. Almost half-million people, many of them Hindu, nearly all unskilled earned their living as street hawkers, street sellers, taxi drivers, mechanics and other manual or industrial trades. Bare-bodied men straining with heavily-laden carts, some pulling from the front, others pushing from behind, up the steps of the bridges in front of the Central Station and near Stanley Hospital, in the scorching mid-day heat.

Some of the handcart pullers would tie a sack-cloth around their feet. In those days, the poor from neighbouring villages migrated to the city in search of work. They came with nothing from their homes - they bathed, ate and slept on the railway platforms, some on the pavements. The whole under light of the most dazzling colours, of mirrored bangles that glistened from rickety market stalls gilded with flags that blew calmly in the warm Indian breeze. Fragrant sacks stacked high in winding narrow alleyways, spice lanes that slapped the air with aromatic flavours never less evocative and never less visceral.

Poor, but here were a people maintained in deeply rooted conviction for the interconnectedness of life and the sanctity of nature. A faith tempered in ceremony, prayer and sacrifice. Their temples blessed and among the holiest place, for the people a way of looking at devotion almost metonymically, a love for the city shouldered on the shape of idols, temples, and shrines. These had, after all, received a visit by the Saivite saints Nayanmars.

The 7th century CE young Saiva poet-saint, Sambandar, would speak of the Region as a place of beautiful groves, with waves that creep up to the shore and then dance on it. As do the fisherfolk who spear the many fish in the waters, and peacocks in its plenty to celebrate the

Thiruvadhirai or the "sacred big wave" using which this universe was created over 132 trillion years in the past. But according to Sambandar, Saiva, perhaps portentously asks "Is it done for you to miss this excitement."

In December and January, a festival for women; unmarried women observe a partial fast on this day to "get good husbands" and married women take a fast from the preceding day and on the day of Thiruvathira for the well-being of their husband and family. In this year, a young 10-year-old Janakiammal from Rajendram, a village close to Marudur Railway Station was such a patnee hona, or a wife-to-be.

The Ramanujan-Janakiammal wedding was a five-day ceremony and it took place along with the wedding of another sister of Janakiammal. They resided at Saiva Muthaiah Mudali Street, in George Town. In this year the 21-yesr-old Ramanujan joined the University of Madras as its first research scholar. His wife and mother lived with him before Ramanujan left for England, on March 17, 1914. Janakiammal joined him in Madras and nursed him till his untimely death on April 26, 1920.

Primes

The theory of primes, as we know, goes back to the Rhind Mathematical Papyrus, from around 1550 BC has Egyptian fraction expansions of different forms for prime and composite numbers. Euclid (c. 300 BC) showed that there are infinitely many primes. Even after 2000 years it stands as an excellent model of reasoning. Here it is:

If the list of primes were finite, then by multiplying them together and adding 1 we would get a new number which is not divisible by any prime on our list - a contradiction.

This is demonstrated trivially;

Suppose p (1), ..., p (n) is a list of all primes, i.e., there are only finitely many. We construct the number N = p (1) x p (2) x...x p (n) + 1.

Viz-a-viz the unique factorization theorem this can be uniquely written as a product of a prime. If p (i) is one of those prime numbers, then N = p (i) x M, and so

1 = N − p (1) x...x p (n) = p (i) x [M − p (1) x...x p (i-1) x p (i+1) x...x p (n)],

which demonstrates that p(i) is a prime divisor of 1. This is, of course, impossible, since p(i) > 1. Thus, none of the prime divisors of N appears in the list of primes. This contradicts the assumption that the list contained all primes. The conclusion is that no such list is complete, and the number of primes must be infinite.

Today, prime numbers are widely studied in the field of number theory. One approach to investigate prime numbers is to study numbers of a certain form. For example, it has been proven that there are infinitely many primes in the form:

a + nd, where d ≥ 2 and GCD (d, a) = 1.

On the other hand, it's still an open question to whether there are infinitely many primes of the form $n^2 + 1$. The Greatest Common Divisor (GCD) of two or more integers, which are not all zero, is the largest positive integer that divides each of the integers. For example, the greatest common divisor of 8 and 12 is 4.

Fermat was first to study numbers (now known as Fermat numbers) in the form $2^{2^n} + 1$ for positive integer n. He conjectured (but admitted he could not prove) all numbers in the form of $2^{2^n} + 1$ to be primes. It then became a question to whether there are infinitely many primes in the form of $2^{2^n} + 1$. The only known Fermat primes are the first five Fermat numbers, $F_0 = 3$, $F_1 = 5$, $F_2 = 17$, $F_3 = 257$, and $F_4 = 65537$.

If $2^k + 1$ is prime, and $k > 0$, it can be shown that k must be a power of 2. (If $k = ab$ where $1 \le a, b \le k$ and b is odd, then $2^k + 1 = (2^a)^b + 1 \equiv (-1)^b + 1 = 0 \pmod{2^a + 1}$ i.e. every prime of the form $2^k + 1$ (other than $2 = 2^0 + 1$) is a Fermat number, and such primes are called Fermat primes. As of 2018, the only known Fermat primes are F_0, F_1, F_2, F_3, and F_4.

The Fermat numbers satisfy the following recurrence relations:

$$F_n = (F_{n-1} - 1)^2 + 1 \text{ for } n \ge 1$$
$$F_n = F_{n-1} - 1^2 + 2^{n-1} F_0 \ldots F_{n-2}$$
$$F_n = F_{n-1} - 1^2 - 2(F_{n-2} - 1)^2$$
$$F_n = F_0 \ldots F_{n-1} + 2$$

for $n \ge 2$. Each of these relations can be proved by mathematical induction. From the last equation, we can deduce Goldbach's theorem 2 Fermat numbers share a common integer factor greater than 1. To see this, suppose that $0 \leq i < j$ and F_i and F_j have a common factor a > 1 then a divides both $F_0 \ldots F_{j-1}$ and F_j hence a divides their difference, 2. Since a > 1, this forces a = 2.

This is a contradiction, because each Fermat number is clearly odd. As a corollary, we obtain another proof of the infinitude of the prime numbers: for each F_n choose a prime factor p_n then the sequence $\{p_n\}$ is an infinite sequence of distinct primes.

Fermat's conjecture was refuted by Leonhard Euler in 1732 when he showed that every factor of F_n must have the form $k2^{n+1} +1$. The fact that 641 is a factor of F_5 can be easily deduced from the equalities $641 = 2^7 \times 5 +1$ and $641 = 2^4 + 5^4$. It follows from the first equality that $2^7 \times 5 \equiv -1 \pmod{641}$ and therefore (raising to the fourth power) that $2^{28} \times 5^4 \equiv 1 \pmod{641}$. On the other hand, the second equality implies that $5^4 \equiv -2^4 \pmod{641}$. These congruences imply that $-2^{32} \equiv 1 \pmod{641}$.

Fermat was probably aware of the form of the factors later proved by Euler, so it seems curious why he failed to follow through on the straightforward calculation to find the factor. One common explanation is that Fermat made a rare computational mistake.

Ramanujan

The young Ramanujan didn't at first jump at the opportunity to work and study with some of the best mathematicians at the turn of the 20th Century. It's entirely possible that his naïve, meek and unimposing assessment of himself prevented him from entirely appreciating the immeasurable opportunity, the gift really, Godfrey Hardy and John

Littlewood had offered him. An invitation, essentially to an ill-educated, jobless unknown to one of England's finest universities. Among Trinity's mathematicians have been John Dee, author of an important Mathematical Preface to the first English translation of Euclid's Elements of Geometry, Isaac Newton who formulated a version of calculus and the principles of universal gravitation and of course Stephen Hawking was a graduate student at Trinity Hall, Cambridge; later, he would hold the title as Lucasian Chair of Mathematics as had Isaac Newton, Joseph Larmor, Charles Babbage, George Stokes, and Paul Dirac. Likewise, Hardy and Ramanujan will be remembered as titans of the College.

Should Ramanujan been anxious to travel - as a strict Brahmin, crossing the seas to foreign lands causes the loss of one's social respectability, as well as the putrefaction of one's cultural character and posterity. It wasn't until early 1914, when Ramanujan gained his mother's consent to make his passage to the Cambridge. But for Ramanujan there were to be more obstacles. He was deeply conflicted by his desire to pursue his mathematics and at the same time conform to the social and religious customs of the times. According to his wife Janakiammal, Ramanujan had sought the blessing of the family's favourite deity, the Goddess Namagiri. And as such, or as a kind of concession he sought Hardy's assurance, modestly, that his diet of fruits, vegetables, beans and grains could be adhered to in England.

Hardy was aware that E.H Neville, another fellow of Trinity College, was [serendipitously] planning to make a trip to Madras at this time. A relieved Ramanujan writes to Hardy in January 1913 – Dear Sir, "I learnt from Mister Neville that you are you're anxious to get me to Cambridge." He goes on, "I went to you colleague of your College, who very kindly spoke to me and cleared my doubts. I need not care for my expenses, that my English is adequate, that I need not to appear for any examination and that I can maintain my vegetarian diet."

Hardy had in fact asked Neville to help Ramanujan secure a scholarship from the University of Madras. Neville writes to the university that "the discovery of the genius of S. Ramanujan of Madras promises to be the most interesting event of our time in the mathematical world …"

Three years later, in April 1919, Ramanujan returns to India, emaciated from prolonged confinement in sanatoria but now with reputation. Questions regarding his diet and England's inclement climate are seen by some as contributing to his ill-health and death in 1920 at the age of 32.

His health was in deed precarious long before he journeyed to England. His health may have been weak due to health conditions in South India and due to genetic factors. In December 1889, Ramanujan contracted Smallpox, though he recovered. His mother bore several children who did not survive infancy. His generally poor health may have contributed.

Empire

European influences were strongest in the towns of India. This was especially true in the old bases of British trade, such as Calcutta, Madras or Bombay, where a new Indian intelligentsia had begun to take root. Whatever the British may have intended, their early rule seems generally to have consolidated the hold of what they regarded as 'traditional' intellectuals, rather than displacing them by new ones, and the authority of Brahmins and of doctrines of caste separation grew stronger.

In the early 1900s Madras was a place, a home to a population that for but the most ignorant observer can deny ominous signs that would soon sweep away coalescence and shape a drama that would unfold in the coming short years and make Mohandas Karamchand Gandhi a household name both in India and across the world. A nation less inclined to meddlesome formality and officious bureaucracy – less inclined to submit to a colonial resolve of a different king, a different time and a different god.

But India was also a nation at once assured of its unfailing patriotism and peasant-born nationalism. Never quite the classic modern nation state, instead a civilisation swathed in deep tradition, religion, people and creatures that held a place of their own in Hindu pantheon as vehicles of gods and goddesses. But this was also a time where a movement, a groundswell that had taken hold in the 1850's continued to inspire the anti-colonial and reformist trends - emerging during the freedom struggle and squarely criticizing the caste system, feudal exploitation, patriarchy and superstition.

In 1887, amongst the temples of Erode stood a modest house on Sarangapani Sannidhi Street in the town of Kumbakonam. A house that would in time become a museum in remembrance to one of the greatest mathematicians of the 20th Century. Srinivasa Ramanujan, literally, "younger brother of Rama", was born on 22

December 1887 into a Tamil Brahmin Iyengar family. In a sad, almost bizarre sense of irony, Ramanujan would be a brother fleetingly. When he was a year and a half old, his mother gave birth to a son, Sadagopan, who died less than three months later. Two years later, his mother gave birth to two more children, in 1891 and 1894, both failing to reach their first birthdays. Like his siblings, Ramanujan will die from poor health at the prime of his life in 1920 in his birth town of Kumbakonam.

Shortly before he died, Hardy's long-time collaborator John Littlewood, mentioned two names; H. F. Baker and E. W. Hobson; both assumed to be the first recipients of Ramanujan's letters. Neither were particularly good choices: Baker worked on algebraic geometry and Hobson on mathematical analysis, both subjects fairly far from what Ramanujan was doing. But in any event, neither of them responded. One can, while perhaps not sympathize with them, can understanding how they might have received his letters and their reaction. For example, in one letter Ramanujan writes about his "discoveries" including the constant:

$$1+2+3.... = -1/12$$

On examination this looks absurd. A divergent sum surely has a positive infinity. At Ramanujan's abatement he very rarely demonstrated derivation. He said for example, an equation means nothing to me unless it expresses a thought of God. It's highly probable that he quite simply didn't see the need, perhaps didn't know better, and/or didn't know how. We know that his lack of formal training contributed greatly to his inability to communicate his ideas. We also know he was influenced by G. S Carr's coaching notes.

G. S Carr's text was in part written to "teach bright English boys for university entrance exams" and not necessarily the best example of demonstrating formal mathematical formulation. This text, made famous by Ramanujan, marked the real starting point of his career. Carr has sections on the obvious subjects, algebra, trigonometry, calculus and analytical geometry, but some sections are developed disproportionally, and particularly the formal side of the integral calculus. The text aside, Ramanujan had exceptional mathematical intuition. He was highly instinctive with abilities to recognize patterns, differentiate and abstract.

Baker and Hobson would have given Ramanujan's letter, which included about 120 hand written theorems, little consideration. It wasn't their field of study and Ramanujan was far from even being an outsider. As we know unpublished or unpublishable untested material is received by established scholars frequently. These arrive from people believing they have hit on an important idea, perhaps a new way of doing things, some even describe their work with revolutionary overtones. Rarely is the science of science achieved in this manner. But Ramanujan, yes, an unknown was soon to be regarded as one of the greatest mathematicians of the twentieth century. Had he, perhaps, stumbled on this theorem independently?

$$1+2+3\ldots = -1/12$$

It's unclear if Baker and Hobson recognised the expression. I don't see how not. Hardy and Littlewood, certainly did. They clearly understood this as special case of Riemann's 1859 zeta (-1) hypothesis. They undoubtedly regarded the Riemann hypothesis as a mathematical holy grail. Even today, the Clay Mathematics Institute regards the Riemann hypothesis one of 7 Millennium Problems attracting a prize of US$1 Million dollars for a correct solution. Finding a proof of the hypothesis is one of the hardest and most important unsolved problems of pure mathematics. Interestingly, the only Millennium Prize problem to have been solved is the Poincaré conjecture, which was solved by the Russian mathematician Grigori Perelman in 2003. He refused to accept the prize.

Riemann was born into a poor Lutheran pastor's family and for all of his short life he was a shy and introverted individual. Like Ramanujan, Riemann had the propensity not to write very much at all. However, the things he did write, the things he discovered were, are important – they have and a revolutionary effect on mathematics. He contributed to many fields of mathematics, such as analysis, geometry, mathematical physics and number theory. He was among the first mathematicians that worked on complex analysis. The kind of geometry he started (Riemannian geometry) is one of the bases of the theory of relativity, developed by Einstein.

The Riemann zeta function $\zeta(s)$ is a function whose argument s may be any complex number other than 1, and whose values are also complex. It has zeros at the negative even integers; that is, $\zeta(s) = 0$ when s is one of -2, -4, -6, ... known as trivial zeros. However, the negative even integers are not the only values for which the zeta function is zero. The other ones are called non-trivial zeros. The Riemann hypothesis is concerned with the locations of these non-trivial zeros.

If proven correct, this would allow a better understanding of how the primes are placed among whole numbers.

Riemann's zeta function for -1, i.e. $\zeta(-1)\zeta = -1/12$

which is only an associative value to the sum of all-natural numbers and requires much number manipulation when analytically continued.

But both Ramanujan and Euler who almost 200 years earlier summed the series to give -1/12 result. Euler's interest was similar to that of Ramanujan: he wanted to see where the rules of mathematics could take him, so he assumed that the sum existed and performed mathematical calisthenics to arrive at the result. Since that time, boundaries have been further drawn, and modern mathematics has become that much more exacting. This has been achieved in part through the establishment of axiomatic properties of numbers.

Incidentally, the −1/12 solution, in lesser hands than that of Ramanujan or Euler is indeed nothing more than a mathematical party trick. Some physicists mistakenly believe that mathematicians have summed the series to produce this result. And some mathematicians mistakenly believe that physicists have summed the series experimentally to produce the solution.

Despite Ramanujan's tendency not to present derivations, in this case he offered two: $1 + 2 + 3 + 4 + \cdots = -1/12$ in chapter 8 of his first notebook. The simpler, less rigorous derivation proceeds in two steps, as follows.

The first key insight is that the series of positive numbers $1 + 2 + 3 + 4 + \cdots$ closely resembles the alternating series $1 -$

$2 + 3 - 4 + \cdots$. The latter series is also divergent, but it is much easier to work with; there are several classical methods that assign it a value, which have been explored since the 18th century.

In order to transform the series $1 + 2 + 3 + 4 + \cdots$ into $1 - 2 + 3 - 4 + \cdots$, one can subtract 4 from the second term, 8 from the fourth term, 12 from the sixth term, and so on. The total amount to be subtracted is $4 + 8 + 12 + 16 + \cdots$, which is 4 times the original series. These relationships can be expressed using algebra. Whatever the "sum" of the series might be, call it $c = 1 + 2 + 3 + 4 + \cdots$. Then multiply this equation by 4 and subtract the second equation from the first:

The boy

At the Kangayan Primary School, Ramanujan performed well. Just before turning 10, in November 1897, he passed his primary examinations in English, Tamil, geography and arithmetic with the best scores in the district. That year, Ramanujan entered Town Higher Secondary School, where he encountered formal mathematics, really for the first time.

By age 11, he had exhausted the mathematical knowledge of two college students who were lodgers at his home. He was later lent a book by S. L. Loney's Advanced Trigonometry. He mastered this by the age of 13 while developing competent theorems on his own. Here now is an example of a young boy relying little on mastery and much on his intuition, instinct.

By 14, he was receiving merit certificates and academic awards that continued throughout his school career, and he assisted the school in the logistics of assigning its 1200 students (each with differing needs) to its 35-odd teachers. He completed mathematical exams in half the allotted time, and showed a familiarity with geometry and infinite series. Ramanujan was shown how to solve cubic equations in 1902; he developed his own method to solve the quartic. The following year, Ramanujan tried to solve the quintic, not knowing that it could not be solved by radicals.

As mentioned, in 1903, when he was 16, Ramanujan obtained from a friend a library copy of *"A Synopsis of Elementary Results in Pure and Applied Mathematics"*, G. S. Carr's collection of 5,000 theorems. Ramanujan reportedly studied the contents of the book in detail. The next year, Ramanujan independently developed and investigated the Bernoulli numbers and calculated the Euler–Mascheroni constant up to 15 decimal places. His peers at the time commented that they "rarely understood him" and "stood in respectful awe" of him.

When he graduated from Town Higher Secondary School in 1904, Ramanujan was awarded the K. Ranganatha Rao prize for mathematics by the school's headmaster, Krishnaswami Iyer. Iyer introduced Ramanujan as an outstanding student who deserved scores higher than the maximum. He received a scholarship to study at Government Arts College, Kumbakonam, but was so intent on mathematics that he could not focus on any other subjects and failed most of them, losing his scholarship in the process.

In August 1905, Ramanujan ran away from home, heading towards Visakhapatnam, and stayed in Rajahmundry for about a month. He later enrolled at Pachaiyappa's College in Madras. There he passed in mathematics, choosing only to attempt questions that appealed to him and leaving the rest unanswered, but performed poorly in other subjects, such as English, physiology and Sanskrit. Ramanujan failed his Fellow of Arts exam in December 1906 and again a year later. Without a FA degree, he left college and continued to pursue independent research in mathematics, living in extreme poverty and often on the brink of starvation.

It was in 1910, after a meeting between the 23-year-old Ramanujan and the founder of the Indian Mathematical Society, V. Ramaswamy Aiyer, that Ramanujan started to get recognition within the mathematics circles of Madras, subsequently leading to his inclusion as a researcher at the University of Madras.

It's at this point and after Hardy receiving the first of Ramanujan's letter petitions Neville to assist Ramanujan secure a scholarship from the University of Madras. Ramanujan even in the stillness of the Madras University library writes "I am already a half-starving man, to preserve my brains, I want food and this is now my first consideration. Any sympathetic letter from you will be helpful to me here to get a scholarship either from the University or from Government."

Neville writes a letter to the university "the discovery of the genius of S. Ramanujan of Madras promises to be the most interesting event of our time in the mathematical world."

The syndicate of Madras University met on April 7, 1910, to discuss if Ramanujan could be taken in as a researcher. While there were dissenting voices, it was then Vice Chancellor P R Sundaram Iyer who prevailed, saying, "Did not the preamble of the act establishing the university specify that one of its functions was to promote research? And, whatever the lapses of Ramanujan's education, was he not a proven quantity as a mathematical researcher?" Three days later, on April 12, the university cleared Ramanujan's scholarship amount of Rs75 a month, along with permission to access the university library and submit his periodical research reports once in three months.

At the same time Ramachandra Rao, a keen mathematician and astronomer who served as President of the Indian Mathematical Society helped Ramanujan secure employment as a clerk at the Madras Port Trust. Here he would ern a salary of only £20 per annum. Attached to his application was a recommendation from E. W. Middlemast, a mathematics professor at the Presidency College. The reference "Ramanujan is a young man of quite exceptional capacity in Mathematics." Three weeks after he had applied, on 1 March, Ramanujan learned that he had been accepted as a Class III, Grade IV accounting clerk, making 30 rupees per month but now with full-time employment.

From the first

Ramanujan's early letters included integrals Hardy himself was working on that he was sufficiently proud of and published so that when seeing the same integrals in Ramanujan's work Hardy immediately realised that here was a mathematician of equiseta taste. In addition, there were formulas here that absolutely stunned him. Hardy says of these, "I have never seen anything least like them before - a single look at them is enough to show that they can only be written by a mathematician of the highest class. They must be true because if they were not true no one would have the imagination to invent them. The writer must be completely honest because great mathematicians are commoner than thieves or humbugs of such incredible skill."

At Trinity College, Hardy and Littlewood thought, having Ramanujan travel to the College would allow them to better make sense of Ramanujan's results, and the fragments of proofs he had provided through correspondence. It wasn't at all clear to them how Ramanujan was arriving with his results. It did seem he was experimenting with numbers, presumably to check if the results gave him an idea of what might be true. It's not clear how he figured out what was actually true - and indeed some of the results he quoted weren't in the end true. But presumably he used some mixture of traditional mathematical proof, calculational evidence, and lots of intuition. But he didn't explain any of this to Hardy and Littlewood.

Instead, he just started conducting a correspondence about the details of the results, and the fragments of proofs he was able to give. Hardy and Littlewood seemed intent on grading his efforts - with Littlewood writing about some result, for example, "(d) is still wrong, of course, rather a howler." Still, they wondered if Ramanujan was "a Euler", or merely "A Jacobi". For anyone with even the slightest familiarity with Euler and Jacobi's work would appreciate instantly just how much regard Hardy and Littlewood had of Ramanujan.

Littlewood continues, "The stuff about primes is wrong" - explaining that Ramanujan incorrectly assumed the Riemann zeta function didn't have complex zeros, even though it actually has an infinite number of them, which are the subject of the whole Riemann hypothesis. Here, it becomes more apparent that Ramanujan had indeed lacked the formal, structured knowledge one obtains through a more formal educative resumé. For example, an understanding of the mathematical phenomena of which a competent mathematician will systematically link and build upon, albeit there are some examples of this as seen in an extract from a January 1913 letter to Hardy below (Fig. 1.) This would include working within a theoretical framework, working within a structure of expert knowledge and with tested mathematical practices.

Just as it cannot be taken for granted that mathematics as a field of study is reflective of mathematics as a cognitive activity it cannot be taken for granted that the ways in which one describes, understands, or even misunderstands an idea with which they have considerable formal experience is indicative of the processes by which that idea was first acquired. In this case Ramanujan is essentially self-taught, unpractised and predisposed to the most difficult of circumstances.

Fig 1. Ramanujan referencing Euler

> Just as in elementary mathematics you give a meaning to a^n when n is negative and fractional to conform to the law which holds when n is a positive integer, similarly the whole of my investigations proceed on giving a meaning to Eulerian Second Integral for all values of n. My friends who have gone through the regular course of University education tell me that $\int_0^\infty x^{n-1} e^{-x} dx = \Gamma(n)$ is true only when n is positive. They say that this integral relation is not true when n is negative. Supposing this is true only for positive values of n and also supposing the definition $n\Gamma(n) = \Gamma(n+1)$ to be universally true, I have given meanings to these integrals and under the conditions I state the integral is true for all values of n negative and fractional. My whole investigations are based upon this and I have been developing this to a remarkable extent so much so that the local mathematicians are not able to understand me in my higher flights.

At Trinity college

Shortly prior to Ramanujan setting sail to London, Littlewood writes to Harding, in an almost exact parallel as when Einstein, says of Paul Dirac, "This balancing on the dizzying path between genius and madness is awful," Littlewood, "How maddening are his letters is in the circumstances I rather suspect that he's afraid you will steel his work." Afterwards, "I received a letter from the War Office this morning, seems they need some assistance with ballistics." And later, he writes to the atheist Hardy. "Please forgive this personal transcreation. I'm gone now to this god-awful war. I haven't the faintest idea if I'll ever return. Fortunately, unlike you I do have god to take comfort in. I have 2 points to make."

1. The Prime number theory is wrong…

2. The other point is less straightforward – you have in Ramanujan nothing short of a miracle. The man exceeds any notions of brilliance that I have ever understood, forget Jacobie, we can compare him to Newton. I have come to believe for Ramanujan every single positive integer is one of his personnel friends. And to that end you too have a responsibility – you have to look after him and make sure his work amounts to something. Don't let Howard and his lot win. So, you see Hardy, you too have a war to fight. Just don't let it be with Ramanujan.

After dinner in Trinity one evening, some of the fellows adjourned to the combination room. Over their claret and port Hardy mentioned to Littlewood some of the claims he had received in the mail from an unknown Indian. Some assertions they knew well, others they could prove, others they could disprove, but many they found not only fascinating and unusual but also impossible to resolve.

This toing and froing between Hardy and Littlewood continued the next day and beyond, and soon they were convinced that their correspondent was a genius. So, Hardy sends an encouraging reply to Ramanujan, which led to a frequent exchange of letters.

It was clear to Hardy that Ramanujan was totally exceptional: however, in spite of his amazing feats in mathematics, he lacked the basic tools of the trade of a professional mathematician. Hardy knew that if Ramanujan was to fulfil his potential, he had to have a solid foundation in mathematics, at least as much as the best Cambridge graduates.

His work on the Partition function with Hardy was the precursor of what is now called the circle method in analytic number theory and has played a huge role in discovers in number theory in the 21st century. The work on probabilistic number theory which has also played a significant activity in the 20th century has its origin in a paper Ram and Hardy developed. The amazing achievements with regard to the theory of Modular forms and applications in number theory, whilst Ramanujan was not the one to discover Modular forms, he nevertheless had an insight in them and an understanding of questions.

Ramanujan claimed, in his letters to Hardy, that he had found a more or less exact formula for the prime counting function $\pi(n)$ Upon closer inspection by Hardy and

Littlewood, this formula was later shown to be incorrect. This is an example of the ongoing tension between mathematical intuition, as exemplified by Ramanujan, and mathematical rigor, as exemplified by Hardy; Hardy emphasizes that situations like this are why intuition is not enough and Ramanujan needs rigor as well. In fact, And Ramanujan said his mathematical formulae came to him as visions provided by a goddess. This way of conceptualising mathematics was anathema to Hardy's code.

In fact, Hardy writes about this error in The Indian Mathematician, Ramanujan's theory of primes was vitiated by his ignorance of the theory of functions of a complex variable. It was (so to say) what the theory might be if the Zeta-function had no complex zeros. His method depended upon a wholesale use of divergent series. That his proofs should have been invalid was only to be expected. But the mistakes went deeper than that, and many of the actual results were false. He had obtained the dominant terms of the classical formulae, although by invalid methods; but none of them are such close approximations as he supposed.

Having said that, Ramanujan does own some primes – a theory about the number of primes between certain points. The *n*th Ramanujan prime p is the smallest prime such that there are at least n primes between x and 2x. This is true for any x such that 2x > p.

The nth Ramanujan prime is the smallest number R_n such that $\pi(x) - \pi(x/2) \geq n$ for all $x \geq R_n$, where $\pi(x)$ is the prime counting function. In other words, there are at least **n** primes between $x/2$ and x whenever $x \geq R_n$. The smallest such number R_n must be prime, since the function $\mu(x) - \mu(x \setminus S)$ can increase only at a prime.

Equivalently,

$$R_n = 1 + \max_k \left\{ k : \pi(k) - \pi\left(\tfrac{1}{2}k\right) = n - 1 \right\}.$$

Using simple properties of the gamma function, Ramanujan (1919) gave a new proof of Bertrand's postulate. Then he proved the generalization that $\pi(x) - \pi(x/2) \geq 1, 2, 3, 4, 5, \ldots$ if $x \geq 2, 11, 17, 29, 41, \ldots$ respectively. These are the first few Ramanujan primes.

The case

$\pi(x) - \pi(x/2) \geq 1$ for all $x \geq 2$ is Bertrand's postulate.

While the prime counting formula tells us how many primes, it isn't practical to list all the possible primes in order to figure every π(x). For example, π(105) = 9592. Although many functions exist that can count these numbers, there isn't one single function that can be called the counting function. All are approximations, i.e. all have margins of error. Two functions from the literature include Minác's formula and Willans' formulae, below.

$$\mu(x) = \sum_{x}^{1-3} \left[\frac{1}{(1-1)i+1} - \left[\frac{1}{(1-1)i} \right] \right]$$

(1) $\qquad \pi(n) = -1 + \sum_{j=1}^{n} F(j),$

where $F(j) = \left\lfloor \cos^2\left(\pi \tfrac{(j-1)!+1}{j}\right) \right\rfloor$.

(2) $\qquad \pi(n) = \sum_{j=2}^{n} H(j),$

where $H(j) = \left(\sin^2\left(\tfrac{\pi}{j}\right)\right)^{-1} \sin^2\left(\pi \tfrac{((j-1)!)^2}{j}\right)$.

In 1917, Ramanujan and Hardy proved that the normal order of the number ω(n) of distinct prime factors of a number n is log(log(n)).

Hardy and Wright give an argument for the conjecture that the number of Fermat primes is finite in their well-known footnote of G. H. Hardy and E. M. Wright, An introduction to the theory of numbers, Oxford University Press, 1979. By the prime number theorem, the probability that a random number n is prime is at most $a/\log(n)$ for some choice of a. So, the expected number of Fermat primes is at most the sum of $a/\log(2^{2^x}+1) < a/2^n$, but this sum is a. However, as Hardy and Wright note, the Fermat numbers do not behave "randomly" in that they are pairwise relatively prime.

But if primes are about multiplication, partitions are about addition. Christian Goldbach wrote to Euler in 1742 with a proposition that amounts to this: every even natural number greater than 2 can be expressed as the sum of two primes. For example, we can write 10 as 5 + 5 or 22 as 3 + 19. What makes this awkward to prove is that primes are about multiplication where's this is about addition.

In number theory and combinatorics, a partition of a positive integer n, also called an integer partition, is a way of writing n as a sum of positive integers. Two sums that differ only in the order of their summands are considered the same partition. (If order matters, the sum becomes a composition.) For example, 4 can be partitioned in five distinct ways.

The theory of partitions of numbers is an interesting branch of number theory. The concept of partitions was given by

Leonard Euler in the 18th century. After Euler though, the theory of partition had been studied and discussed by many other prominent mathematicians like Gauss, Jacobi, Schur, McMahon, and Andrews etc. but the joint work of Ramanujan with Hardy made a revolutionary change in the field of partition theory of numbers. Ramanujan and Hardy invented circle method which gave the first approximations of the partition of numbers beyond 200.

The value of $p(7)$ is 15, the partitions being displayed in the grid on the right. Note that they are 'unordered', in the sense that 1 + 6, say, is regarded as the same as 6 +. 1 and is omitted. By n = 70, the number has risen to over 4 million; the Hardy-Ramanujan asymptotic estimate closely shadows this growth (the graphs below centre and right). Exactly how closely, is seen by taking the ratio $p(n)$/H-R (bot- tom right), the lower curve meeting the line y = 1 at infinity.

Ramanujan had already made deep discoveries about the partition function when he was himself discovered' by Hardy in 1913. Five years later, despite Ramanujan being already terminally ill, they published this remarkable asymptotic formula, derived from a series which yields the *exact value* of $p(n)$ in terms $\sqrt{n}$. It was discovered independently, in 1920, by James Victor Uspensky. The Asymptotic Partition Formula looks like this:

$$P(n) \sim \frac{1}{4n\sqrt{3}} e^{\pi \sqrt{2n/3}}$$

The proof of the asymptotic formula for the partition function given by Hardy and Ramanujan was "the birth of the circle method", and used properties of modular forms.

Erdös was trying to give a proof with elementary methods. He succeeded in 1942 to give such a proof, but only with "unknown" constant a. Afterwards Newman gave a "simplified proof", and obtained also that:

$$a = \frac{1}{4n\sqrt{3}}$$

The notebooks

Ramanujan's lost notebook is the manuscript in which Ramanujan recorded the mathematical discoveries of the last year (1919 – 1920) of his life. Its whereabouts were unknown to all but a few mathematicians until it was rediscovered by George Andrews in 1976, in a box of effects of G. N. Watson stored at the Wren Library at Trinity College, Cambridge. In April 2011, Andrews presented a talk (MillerComm 2911) which explored amongst other things whether social or religious or ethnic limitations impact the validity of science and mathematical discovery.

Various aspects of the relationship between religion and science have been addressed by modern historians of science and religion, philosophers, theologians, scientists, and others from various geographical regions and cultures.

From thinking about Ramanujan, Andrews concluded that there are no social, religious or ethnic limitations on the validity of what a scientist discovers. The discovery of a scientists of one civilisation or nationality can be received, assessed and assimilated by scientists of any other civilisation or nationality assuming of course that the receptive scientist is sufficiently informed regarding the state of a subject and has the intelligence and scientific training to comprehend what is offered to him.

The mathematics which Ramanujan did in India can be assessed be British of the highest order to the extent to that they can retrace the steps to which his initiative powers had enabled him to leap over. It was not the Indian-ness of the Ramanujan mathematics which baffled the first British mathematicians Hopson and Baker whom he approached, rather it was their exceptionally advanced originality. It required two mathematicians of very high quality – Hardy and Littlewood to appreciate, to learn from and contribute to Ramanujan's work.

As the years passed and his notebooks have been studied his mathematics has been interpreted, proved and assimilated by Western mathematicians. I've not read any references to the specifically Indian character of Ramanujan's mathematics – his mathematics are mathematics indifferently of the of the place and circumstance of their creation. The intimate reciprocal relationship between genius and tradition are evident in the case of Ramanujan. He made a connection of with some older and incomplete condensations of the traditions of mathematics developed in Europe.

He re-traced in his own way and in ignorance of them paths of the tradition which had already traversed in Europe. In other respect he shot well ahead of the points reached by the movement and tradition of Europe. It is unlikely that the tradition of the Hindu belief in which Ramanujan participated steadfastly obstructed his advances from the traditions of mathematics which had developed in Europe. Ramanujan thought that the Hindu

Gods in whom he believed had in fact inspired his mathematical advances.

How culturally tied down is mathematics and what sorts of responsibilities we should make to that. There are in evidence in the world of education explicit statements by people who are highly placed and knowledgeable that differences in culture truly do affect both mathematics and how it should be taught and what should go on.

3: The Zonda and the sun

The Zonda and the Sun are quarrelling over who is a more powerful force. Suddenly they see a traveller coming down the roadway. They decide; whoever can get the traveller to take off his coat is the winner. The Zonda blows as hard as it can, but the traveller tightens his coat up even harder. Then the humble Sun softly shines its rays, in a kind, gentle manner. The traveller begins to feel the warmth and finally removes his coat with pleasure.

Princeton, 20 March 1956

Dear Mr. von Neumann:

With the greatest sorrow I have learned of your illness. The news came to me as quite unexpected. Morgenstern already last summer told me of a bout of weakness you once had, but at that time he thought that this was not of any greater significance. As I hear, in the last months you have undergone a radical treatment and I am happy that this treatment was successful as desired, and that you are now doing better. I hope and wish for you that your condition will soon improve even more and that the newest medical discoveries, if possible, will lead to a complete recovery.

Since you now, as I hear, are feeling stronger, I would like to allow myself to write you about a mathematical problem, of which *your opinion would very much interest me: One can obviously easily construct a Turing machine, which for every formula F in first order predicate logic and every natural number n, allows one to decide if there is a proof of F of length n (length = number of symbols). Let ψ(F,n) be the number of steps the machine requires for this and let φ(n) = maxF ψ(F,n). The question is how fast φ(n) grows for an optimal machine. One can show that φ(n) ≥ k · n. If there really were a machine with φ(n) ~ k · n (or even ~ k · n²), this would have consequences of the greatest importance. Namely, it would obviously mean that in spite of the undecidability of the Entscheidungsproblem, the mental work of a mathematician concerning Yes-or-No questions could be completely replaced by a machine.*

After all, one would simply have to choose the natural number n so large that when the machine does not deliver a result, it makes no sense to think more about the problem. Now it seems to me, however, to be completely within the realm of possibility that φ(n) grows that slowly. Since it seems that φ(n) ≥ k · n is the only estimation which one can obtain by a generalization of the proof of the undecidability of the Entscheidungsproblem and after all φ(n) ~ k · n (or ~ k · n²) only means that the number of steps as opposed to trial and error can be reduced from N to log N (or (log N)²). However, such strong reductions appear in other finite problems, for example in the computation of the quadratic residue symbol using repeated application of the law of reciprocity. It would be interesting to know, for instance, the situation concerning the determination of primality of a number and how strongly in general the number of steps in finite combinatorial problems can be reduced with respect to simple exhaustive search.

I do not know if you have heard that "Post's problem", whether there are degrees of un-solvability among problems of the form (∃ y) φ(y,x), where φ is recursive, has been solved in the positive sense by a very young man by the name of Richard Friedberg. The solution is very elegant. Unfortunately, Friedberg does not intend to study mathematics, but rather medicine (apparently under the influence of his father). By the way, what do you think of the attempts to build the foundations of analysis on ramified type theory, which have recently gained momentum? You are probably aware that Paul Lorenzen has pushed ahead with this approach to the theory of Lebesgue measure. However, I believe that in important parts of analysis non-eliminable impredicative proof methods do appear.

I would be very happy to hear something from you personally. Please let me know if there is something that I can do for you. With my best greetings and wishes, as well to your wife.

*Sincerely yours,
Kurt Gödel*

P.S. I heartily congratulate you on the award that the American government has given to you.

Like most, writing about Kurt Gödel necessitates careful choice between the man's math, his work in logic, his philosophical views, his psychiatric condition, as well as braving his solution of Einstein's equations describing a rotating universe – presented on the occasion of Einstein's 70th birthday. Each of these have been well stayed by plentiful authors and scholars but yet Gödel remains, far from any other mathematician least understood.

The Austrian logician, brilliant, however beset by misfortunes of all kinds, the attention of the biographers with mixed results and mixed worth. One wonders about the rights of the deceased in such matters. These lines from Othello's final soliloquy (from Act 5 of Shakespeare's play) seem quite apt: I have done the state some service and they know't. No more of that. I pray you, in your letters, when you shall these unlucky deeds relate, Speak of me as I am. Nothing extenuate, Nor aught set down in malice.

As we shall see, Gödel was extremely upset at how his work was received, the negative and erroneous interpretation, spread far and wide, that somehow certain mathematical questions would never be resolved, the propaganda that he somehow proved.

Verena Huber-Dyson on Gödel: "he had a fantastic sense of humour, something else that is not written about or talked about much. It is a shame, really." The Swiss-American mathematician, known for work in group theory and formal logic goes on to say; "I think, about all the crazy rumours and gossip about him, simply leave the man (who is now dead) alone!" "He was a sweet and kind man...." This is something that impressed me more than his mind although his voice is more unusual than being kind - and Gödel's voice was simply nothing that existed in this world. He spoke from another world."

The late John Wheeler might not have completely agreed with the "sweet natured" portrayal of Gödel as he reports having been "thrown out" (figuratively, surely) of Gödel's office for suggesting similarities between Incompleteness and Heisenberg's undecidability. "I chalked it up to Gödel having been brainwashed by Einstein," barks Wheeler with all the joy of friendship, camaraderie and banter. The relationship between Gödel and Einstein was indeed special. Gödel used to walk every day with his friend Einstein at the Institute of Advanced Study at Princeton. Einstein told a colleague that in the later years of his life, his own work, which had married space to time and spawned the atom bomb, no longer meant much to him and that he used to come to the institute merely "to have the privilege to be able to walk home with Gödel."

Gödel's work is often supposed inaccessible to all but for the professional mathematician, a grounding in symbolic logic, and number theory bodes well. From the outside we associate Gödel's work with deeply rooted, highly esoteric insights. But, from time to time, beguiling headlines find their way into popular culture and media. Examples include; "Gödel's ontological proof is an argument for God's existence" or as mentioned above, "Gödel makes time travel a possibility." Despite these, I think Gödel's work can be understood, if perhaps not always Gödel himself.

Across the years, in particular after his death in 1978 copious literature describe Gödel's mental anguish and personal tragedy, something, I as well would want to touch upon. Off course his drifting mental state made a mark on his professional work, on his personal and professional relationships as well as obviously upon himself. At Princeton, for example, between 1940 and 1944 he would be offered only temporary appointments, to be renewed each year. Only in 1946 was he given a permanent appointment – still as a visitor, not as one of the permanent faculty. In 1953, at age forty-seven, he was finally made a professor – and even that came after he had been awarded honorary degrees by Yale and Harvard and shared the first Einstein Medal in 1951. Doubtless the delay was due to considerable apprehension on the part of several facility members worried about his mental health.

The character and achievement of Gödel, the most important logician since Aristotle, bear comparison with those of the most eminent mathematicians. His aversion to controversy is reminiscent of Newton's, while his relatively small number of publications - each quite precise and almost all making a major contribution, echo Gauss's motto of "few but ripe." Like Newton, he revolutionized a branch of mathematics in this case, mathematical logic giving it a structure and a Kuhnian paradigm for research.

While it's true Gödel was by no means a prolific writer this point has often been mistaken with a certain belief he had an aversion to writing – specially personnel in nature. Claims that Gödel rarely wrote, or that he didn't write except in formal logic are in the main much exaggerated.

I think it's somewhat of an ignominy that more of his less technical writings haven't been sought and haven't been persevered. Gödel did write and he wrote with warmth, kindness, sincerity and care. Obviously, translation from the German may have presented some source of emotional forfeiture but then we only have to look at how wonderfully Johann Wolfgang von Goethe's works have come across to English. It would be difficult for any reputable scholar to botch the essence of the man's thoughts nuanced free of rigid logic.

Those familiar with Gödel's work will recognise the letter. It's an important letter for reasons, not least because of its simplicity, smooth, and beautiful style. You will know this to be Gödel's lost letter or simply Gödel's letter.

It's an important letter for far other reasons, more so, for the fact that the problem Gödel is referring to continues to be one of the most important questions in computability and mathematics. The lost letter is hence asking whether a certain NP complete problem (NP stands for nondeterministic polynomial time) could be solved in linear time. In computably theory, there is an open problem - can every solved problem whose answer can be checked quickly by a computer also be quickly solved by a computer? P and NP are the two types of maths problems referred to: P problems are fast for computers to solve, and so are considered "easy". NP problems are fast (and so "easy") for a computer to check, but are not necessarily easy to solve.

It currently appears that P ≠ NP, meaning we have plenty of examples of problems that we can quickly verify potential answers to, but that we can't solve quickly. All of these problems share a common characteristic; that is the key to understanding the intrigue of P versus NP: In order to solve them you have to try all possible combinations. Here are two examples.

Example one

A traveling salesman wants to visit 100 different cities by driving, starting and ending his trip at home. He has a limited supply of gasoline, so he can only drive a total of 10,000 Km. He wants to know if he can visit all of the cities without running out of gasoline.

Example two

A farmer wants to take 100 watermelons of different masses to the market. She needs to pack the watermelons into boxes. Each box can only hold 20 Kgs without breaking. The farmer needs to know if 10 boxes will be enough for her to carry all 100 watermelons to market.

What is also noteworthy is Gödel's reference to von Neumann's illness. von Neumann spent his last year in a hospital, dying from cancer he developed as a result of his work on the atomic bomb. Not only was Von Neumann only periodically lucid during these last months the detail of his illness and treatment was not readily made public even amongst those close to him. How Gödel would have known about this treatment at the time he supposedly wrote the letter remains contentious. I very much believe Gödel wrote the letter.

In another quirk of history, when this letter first came to light in 1988, no one knew exactly who to credit the P≠NP problem. In the 1980's the P≠NP problem was more or less known as the Cook-Levin Conjecture. To add to this, Richard Karp, in 1971 paper "Reducibility Among Combinatorial Problems" used Stephen Cook's work to develop his 21 NP-complete problems. The contribution Karp made was manyfold: he restated Cook's theorem, he improved the theorem, he named the problem, and most importantly, and he showed how this problem had a wider consequence than Gödel or Cook realized.

Gödel died on January, 14, 1978 of malnutrition, weighing only 65 pounds. He refused food until it was first tasted by his wife, Adele Nimbursky. However, when she became ill in 1977 and had to be hospitalized for six months, Gödel simply refused to eat, effectively starving himself to death.

Gödel had suffered from poor health most of his life. Beginning with an episode of rheumatic fever at the age of six, he remained convinced he had never fully recovered and was known for being paranoid, anxious, and depressed. At age eight, after reading a medical book, Gödel became convinced that he had a weak heart, a possible complication of the rheumatic fever that he had recovered from when he was six. Nevertheless, hypochondriacal concerns would become a lifelong preoccupation. He suffered several nervous breakdowns throughout his life - his mental health never seemed to stabilize. Towards the end of his life, Gödel's paranoia only grew - his death certificate reported that he died of "malnutrition and inanition caused by personality disturbance."

Half a century earlier, at 18 years of age, he began his intellectual career studying theoretical physics, mathematics, and philosophy at the University of Vienna. 1920s Vienna was a thriving intellectual metropolis, and he was surrounded by a group of well-known thinkers, mainly philosophers that composed the famous Der Wiener Kreis (the Vienna circle.) Here, the group would discuss foundational problems, many members inspired by Ludwig Wittgenstein's Tractatus Logico-Philosophicus published in German in 1921.

The challenge, to identify, verify and clarify the relationship between language and reality and to define the limits of science, a theory of knowledge which asserted that only statements verifiable through empirical observation are cognitively meaningful, tenets that prove instrumental in Gödel's future work - something that Gödel would address with prodigious mental force. In 1931, Gödel at the age of twenty-five demonstrated, in effect, that hopes of reducing mathematics to an axiomatic system, as envisioned by mathematicians and philosophers at the turn of the 20th century was in vain - is in vain.

Gödel began his 1951 Gibbs Lecture by stating "research in the foundations of mathematics during the past few decades has produced some results which seem to me of interest, not only in themselves, but also with regard to their implications for the traditional philosophical problems about the nature of mathematics." Gödel is referring here especially to his own incompleteness theorems.

Gödel published proofs of the relative consistency of the axiom of choice and the generalized continuum hypothesis in 1938, his findings strongly influenced the (later) discovery that a computer can never be programmed to answer all mathematical questions, in other words, the P ≠ NP problem.

But despite all, and not withstanding Gödel's dissertation adviser, Hans Hahn, was one of the leaders of the Vienna Circle, Gödel never felt at ease with the Circle, he didn't fit in, since his theistic beliefs clashed with the popular ideas of logical positivism; arguing that the only real knowledge is that which can be demonstrated empirically.

Fitting in the Circle, by all accounts, came secondary to a more desperate, a more insidious set of events - the many Jewish members of the Vienna Circle ensured the group would be regarded with suspicion by the Nazis. Other members were tainted by association. Mathematical logic and set theory were "Jewish mathematics" according to the Nazi orthodoxy. In 1936, Mortiz Schlick, the founder of the Vienna Circle, and the man whose work had first interested Gödel in mathematical logic was murdered by Johann Nelböck, a former student of Schlick and a Nazi sympathiser. Just how this would mould Gödel is for psychiatrists to assess. These events, nevertheless, are said to have been a trigger, the onset to his psychiatric illness from which he would never overcome.

In his philosophical work Gödel would formulate and defend mathematical Platonism - the idea that mathematics is a descriptive science, or alternatively the idea that the concept of mathematical truth is objective; Propositions are about the relation of concepts. Accordingly, mathematics does not create new objects but, instead, discovers already existing objects. These exist in the non-material world of abstract ideas, which is accessible only to our intellect - a true mathematical proposition is therefore true in virtue of meaning. On that basis Gödel laid the foundation for the program of conceptual analysis within set theory. He adhered to Hilbert's "original rationalistic conception" in mathematics (as he called it); and he was prophetic in anticipating and emphasizing the importance of large cardinals in set theory before their importance became clear.

Gödel's philosophical ideals would evolve and coalesce, but generally, as a discipline of fundamental knowledge, reality and existence, his philosophy would give his mathematical theorems a more universal underpinning. Many years later, in the late 1970s, when Gödel reflected upon his career, he told to Hao Wang, who chronicled Gödel's philosophical ideas and authored several books on the subject; "when I entered the field of logic, there were fifty percent philosophy and fifty percent mathematics. There are now ninety percent mathematics and only one percent philosophy." His views broadly characterized by two points of focus, or, in modern parlance, commitments. Realism, the belief that mathematics is a descriptive science and, a form of

Leibnizian rationalism in philosophy; and in fact, Gödel's principal philosophical influences, in this regard particularly but also many others, were Leibniz, Kant and Husserl.

The Leibnizian dialogue, at the heart and root of the philosophy is a connecting role of the dialogue within the Leibnizian thought as explained on a number of levels. From the metaphysical perspective the possibilities and limits of the dialogue in the horizon of human knowledge and the level or the function of dialogue in the dynamics of the human invention. To assess the importance and influence this had on Gödel, one need only to consider the following:

> Gödel - "...our total reality and total experience are beautiful and meaningful, we should judge reality by the little which we truly know of it. Since that part which conceptually we know fully turns out to be so beautiful, the real world of which we know so little should also be beautiful."

Gödel's realism and Gödel's rationalism must be prefaced with a disclaimer: there is no single view one could associate with each of these. As noted, Gödel's realism underwent a complex development over time, in both the nature of its ontological claims as well as in Gödel's level of commitment to those claims. Similarly, Gödel's rationalism underwent a complex development over time, from a tentative version of it at the beginning, to what was adjudged to be a fairly strong version of it in the 1950's. Around 1959 and for some time afterward Gödel fused his rationalistic program of developing exact philosophy with the phenomenological method as developed by Husserl.

Gödel published his views on realism for the first time in his 1944. The following is one of his most quoted passages on the subject:

> "...Classes and concepts may, however, also be conceived as real objects, namely classes as "pluralities of things," or as structures consisting of a plurality of things and concepts as the properties and relations of things existing independently of our definitions and constructions..."

and

> "Mathematical concepts, not mathematical objects, are what mathematics is about"

These were formulated predominantly in the context of the foundations of mathematics and set theory. Although the roots of Gödel's belief in rationalism are metaphysical in nature, his long-standing aspirations in that domain had always been practical ones; namely, to develop exact methods in philosophy; to transform it into an exact science, or "strenge Wissenschaft" (rigorous science) to use the German philosopher, Husserl's term. After the 1960's Gödel's realist views were formulated mostly in the context of the foundations of mathematics and set theory.

The judgement levied upon Gödel's rationalism by contemporary philosophers was, however, a harsh one. Nevertheless, Gödel himself remained optimistic as seen by his comments to Wang: "it is not appropriate to say that philosophy as rigorous science is not realizable in the foreseeable future. Time is not the main factor; it can happen anytime when the right idea appears."

Reconciling realism and rationalism presents a conundrum. The high standard imposed on philosophical arguments by rationalism demands something like a proof of the realist position. But it is far from clear how to go about proving the existence of an acausal, atemporal domain of mathematical objects, especially with the notion of mathematical proof in mind. As we saw, Gödel observed in the Gibbs lecture that a proof would amount to giving an exhaustive enumeration of the alternatives to realism, and then refuting them one by one.

But this dialectical approach to the problem of proving the realist view, "the only one tenable", did not lead to its solution. Regarding his other major philosophical work from the 1950s, after six years of work on it, Gödel notified the editors in 1959, that he would not be submitting his main philosophical work from the period 1953 to 1959, the fundamental paper "Is Mathematics a Syntax of Language?"

In 2013, the German conservative newspaper Die Welt reported "Scientists Prove Existence of God," As an eye-catching headline as it was, unsurprisingly, there is a rather significant caveat to that claim. When Gödel died in 1978, he left behind a tantalizing theory based on principles of modal logic - that a higher being must exist. The details of the mathematics involved in Gödel's ontological proof are complicated, but in essence he was arguing that, by definition, God is that for which no greater can be conceived. And while God exists in the understanding of the concept, we could conceive of him as greater if he existed in reality. Therefore, he must exist.

Even at the time, the argument was not exactly a new one. For centuries many have tried to use this kind of abstract reasoning to prove the possibility or necessity of the existence of God. But the mathematical model composed by Gödel proposed a proof of the idea. It's theorems and axioms, assumptions which cannot be proven can be expressed as mathematical equations, and that means they can be proven. The Die Welt piece told the story of how scientists using an ordinary MacBook computer showed that Gödel's proof was correct - at least on a mathematical level - by way of higher modal logic.

Ultimately, the formalization of Gödel's ontological proof is unlikely to win over many atheists, nor is it likely to comfort true believers, who might argue the idea of a higher power is one that defies logic by definition. For mathematicians looking for ways to break new ground, however, the news could represent an answer to their prayers.

As mentioned, the judgement levied upon Gödel's rationalism by contemporary philosophers was a harsh one. Gödel remained optimistic as gleaned from the comments he made to Wang: "It is not appropriate to say that philosophy as rigorous science is not realizable in the foreseeable future. Time is not the main factor; it can happen anytime when the right idea appears."

One of the main advances at the intersection of mathematics and philosophy is the development of foundations of mathematics in the early 20th century (type theory and set theory). Intra-mathematical considerations of course played a very important role in motivating such developments, but philosophical concerns shaped and drove much of this development. The same is true for the subsequent development of disciplines of mathematical logic such as proof theory and computability theory.

The nature of truth is a central topic in metaphysics and philosophy of logic, and work on truth is closely connected as already discussed to epistemology and philosophy of language. Significant advances have been achieved in formal theories of truth, and there are close connections between philosophical work on truth and model theory (especially of arithmetic.)

The discovery of non-Euclidean geometries (in the nineteenth century) undermined the claim that Euclidean geometry is the one true geometry and instead led to a plurality of geometries no one of which could be said (without qualification) to be "truer" than the others. In a similar spirit many have claimed that the discovery of independence results for arithmetic and set theory (in the twentieth century) has undermined the claim that there is one true arithmetic or set theory and that instead we are left with a plurality of systems no one of which can be said to be "truer" than the others.

When Gödel suggested that it wasn't appropriate to say that philosophy as rigorous science is not realizable in the foreseeable future he was well versed in the ideas of mathematical theories about mathematical theories. The German mathematician, David Hilbert's work, which come to be known as Hilbert's Program attempted to secure the foundations of mathematics in the early part of the 20th century. It calls for a formalization of all of mathematics in axiomatic form, together with a proof that this axiomatization of mathematics is consistent. He would write "metamathematics provides a rigorous mathematical technique for investigating a great variety of foundation problems for mathematics and logic."

Fig 1.

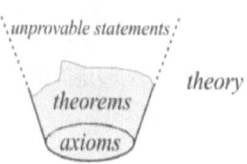

When using the axiomatic method, the treatment of the field of interest commences by carefully selecting a few *basic notions* and making a few basic statements, called *axioms* (see Fig. 1). An axiom is a statement that asserts either that a basic notion has a certain property, or that a certain relation holds between certain basic notions. The basic notions and axioms form our *initial theory* of the field.

An important feature of metamathematics is its emphasis on differentiating between reasoning from inside a system and from outside a system. An informal illustration of this is categorizing the proposition "2 + 2 = 4" as belonging to mathematics while categorizing the proposition "'2 + 2 = 4' is valid" as belonging to metamathematics.

But even before Hilbert, in 1868, Eugenio Beltrami proved that Euclid's fifth axiom cannot be deduced from the other four axioms. In two-dimensional geometry a line parallel to a given line L is a line that does not intersect with L. Euclid's fifth axiom, also called the Parallel Postulate, states that at most one parallel can be drawn through any point not on L.

To Euclid and other ancients, the fifth axiom seemed less obvious than the other four of Euclid's axioms. This is because a parallel line can be viewed as a line segment that never intersects with L, even if it is extended indefinitely. The fifth axiom thus implicitly speaks about a certain occurrence in arbitrarily removed regions of the plane, that is, that the segment and L will never meet. However, since Aristotle the ancients were well aware that one has to be careful when dealing with infinity. For example, they were already familiar with the notion of asymptote, a line that "approaches a given curve but it meets the curve only in the infinity." To avoid the vagueness and controversy of Euclid's fifth axiom, they undertook to deduce it from Euclid's other four axioms; these caused no disputes. However, all attempts were unsuccessful until 1868.

The importance of Beltrami's discovery is that it does not belong to geometry, but rather to the science about geometry, and, more generally, to metamathematics, the science about mathematics. The realization that instinct and experience can be delusive, mathematics gradually took a more abstract view of its research subjects. And, geometry this was just the start. Set theory, a branch of mathematics that deals with the properties of well-defined collections of objects, made a mathematical discipline by the German mathematician and logician Georg Cantor followed shortly after. In fact, such was the trouble set theory was to find itself into that that the discipline was in the frame of the so-called naive set theory.

In the early 20th century, among growing concerns about various foundational issues, it was noticed that there are a number of paradoxes that can be produced by infinite sets. Throughout the 20th century, mathematicians developed various axiomatic systems to make set theory more rigorous and to free it from these paradoxes. Cesare Burali-Forti demonstrated that constructing "the set of all ordinal numbers" leads to a contradiction and therefore shows an antinomy in a system that allows its construction. And, Cantor himself discovered a similar paradox in 1899.

The British philosopher, logician, mathematician, and historian Bertrand Russel showed in 1903 that some attempted formalizations of the naïve set theory created by Cantor led to a contradiction. Russell's paradox is the most famous of the logical or set-theoretical paradoxes. Also known as the Russell-Zermelo paradox, the paradox arises within naïve set theory by considering the set of all sets that are not members of themselves. Such a set appears to be a member of itself if and only if it is not a member of itself. More formally; he found that in Cantor's set theory there exists a set R that both is and is not a member of itself.

> Let the set R be defined by;
>
> $R = \{S \mid S \text{ is a set} \land S \text{ does not contain itself as a member}\}$ must exist because of the *Axiom of Abstraction*.
>
> The *Law of Excluded Middle* guarantees that R either contains itself as a member (i.e., $R \in R$), or does not contain itself as a member (i.e., $R \notin R$).
>
> But then, using the definition of R, each of the two alternatives implies the other, that is, $R \in R \iff R \notin R$. Hence, each of the two is both a true and a false statement in Cantor's set theory.

In English, the Barber paradox is often suggested as an alternative form of Russell's paradox, however, Russell, himself denied that the paradox was an instance of his own.

> *The barber is the "one who shaves all those, and those only, who do not shave themselves." The question is, does the barber shave himself?*

Answering this question results in a contradiction. The barber cannot shave himself as he only shaves those who do not shave themselves. As such, if he shaves himself he ceases to be the barber. Conversely, if the barber does not shave himself, then he fits into the group of people who would be shaved by the barber, and thus, as the barber, he must shave himself.

We could ask why we fear paradoxes? Suppose, that a theory contains a logical statement such that both the statement and its negation can be deduced. Then it can be shown that *any* other statement of the theory can be deduced as well. So, in this theory everything is deducible! This, however, is not as good as it may seem at first glance. Since deduction is a means of discovering truth (i.e., what is deduced is accepted as true) we see that in such a theory every statement is true. But a theory in which everything is true has no cognitive value and is of no use. Such a theory must be cast off.

By 1929, some sort of incompleteness phenomenon was not totally unexpected. After all, mathematicians occasionally considered the question of impossibility in the context of a several fundamental problems of arithmetic, geometry, and algebra supposedly, as far back as in about 550BC, when the Pythagoreans first encountered the 'irrationality' of numbers like √2 which cannot be expressed as the ratio of two integers.

The possibility of incompleteness in the context of set theory was discussed by Bernays and Tarski already in 1928, and von Neumann, in contrast to the dominant spirit in Hilbert's program, had considered it possible that logic and mathematics were not decidable. Gödel was also impressed by Brouwer, who in his lecture in Vienna in 1928 had suggested that mathematics is inexhaustible and cannot be completely formalized.

Hilbert's programme began confidently and he believed that it would just be a matter of time before all of mathematics was corralled within its formalistic web. Alas, the world was soon turned upon its head by the young Gödel. Gödel had completed one of the early steps in Hilbert's programme as part of his doctoral thesis, by proving the consistency and completeness of 1st order logic. But the next steps that he took have ensured his fame. Far from extending Hilbert's programme to achieve its key objective, a proof of the completeness of arithmetic, Gödel proved that any system rich enough to contain arithmetic must be incomplete and undecidable.

By now, the case for mathematical ideas unhindered with contradictions was well made; ideas should not be true and false at the same time. A system without contradictions is *consistent*. Gödel said that *every non-trivial (interesting) formal system is either incomplete or inconsistent*.

1. There will always be questions that can't be answered, using a certain set of axioms;
2. You can't prove that a system of axioms is consistent, unless you use a different set of axioms.

Those theorems are important to mathematicians because they prove that is impossible to create a set of axioms that explains everything in maths. His completeness theorem of 1929 proved that any logical truth that can be stated in the predicate calculus is provable by the axioms of deduction. His incompleteness theorem of 1931, in contrast, proved that no finitely or recursively enumerable axiom system that is consistent, and strong enough to include arithmetic, can be complete: there must exist true theorems in arithmetic that cannot be proved in it. In a second incompleteness theorem proved shortly afterwards he showed that the consistency of such a theory is also unprovable.

As already mentioned, the intellectual environment of Gödel in the 1930s was that of the Circle with its radically anti-metaphysical attitude. In particular, even the notion of truth was considered as suspicious or even nonsensical at the time, at least by some logical positivists. Therefore, Gödel worked hard to eliminate any appeal to the notion of truth and attempted to do without it. He therefore introduced the notion of ω-consistency, which can be defined rigorously and purely syntactically. This led to the incompleteness theorems in the form that they are now known.

After completing secondary school at Brno, in 1924 Gödel went to Vienna to study physics at the university. The elegant lectures of P. Furtwangler (a pupil of Hilbert and cousin of the famous conductor) fed his interest in mathematics, which became his major area of study in 1926. His principal teacher, as mentioned, was the analyst Hans Hahn, someone actively interested in the foundations of mathematics.

It's interesting that he would choose mathematics – back in February 10, 1917, Gödel, a very young child, in a typical example of the old-fashioned rote method of teaching, he would write the symbols "+" and the numbers "2" and "Y' a hundred (or more) times, presumably for practice, and then finally the equations "1 + 1 = 2" and "2 + 1= 3." But then his mind wandered and he wrote "1+ = 21." His report card for that year of rated the 10-year-old "Sehr gut" (Very well) in every subject except mathematics, in which he only rated a "Gut" (Well.)

Gödel began his studies in Vienna in 1924 in physics with Hans Thirring; and although he switched to mathematics two years later, studying first number theory under Furtwingler and then logic under Hahn and, informally, Carnap, he continued to take physics courses up to the time he graduated in 1929. The background in physics Gödel obtained as a student may explain his contribution to Relativity in 1947.

In a letter to Seelig (Einstein's biographer) in 1955, Gödel writes "I have, however, in connection with certain philosophical problems, devoted myself for some time to a less difficult complex of questions from general relativity theory, namely cosmology. The fact that here I, as a newcomer to the field of relativity theory, could immediately obtain essentially new results seems to me sufficient to demonstrate the unfinished state of the theory."

Gödel's doctoral dissertation, "Über die Vollständigkeit des Logikkalküls" ("On the Completeness of the Calculus of Logic") was completed in the autumn of 1929, and he received the Ph.D. on February 6, 1930. A somewhat revised version was presented at Karl Menger's colloquium on May 14, 1930, and was published (1930a) using "several valuable suggestions" of Professor Hahn "regarding the formulation for the publication."

The 1930s were a prodigious decade for Gödel. After publishing his 1929 dissertation in 1930, he published his ground-breaking incompleteness theorems in 1931, on the basis of which he was granted his Habilitation (qualification to conduct self-contained university teaching and is the key for access to a professorship) in 1932 and a Privatdozentur at the University of Vienna in 1933.

The Completeness theorem states that: If τ is a first-order-sentence such that τ is valid (true under any interpretation), then τ is provable from the axiomatic frame of the first order logic. To understand this, it's helpful to remember that while studying logic, we make a distinction between the syntactic and the semantic perspectives. we deal with the syntax and semantics separately and then tie them together by soundness and completeness theorems.

In the syntax, you have some set of formulas called "logical axioms" which is a fixed set. and a set of sentences called "non-logical axioms" which varies according to the topic you have in mind and some "rules of inference " which tells you which sentences you can deduce from which.

You have to know that, non-logical axioms vary as mentioned, for example, if you discuss group theory then you will pick some axioms different from those of topological spaces. So, non-logical axioms determine the theory you wish to work on. logical axioms represent the axioms that are Always valid i.e. no matter which subject you work on, they are always satisfied, logical axioms could be something like (a = b) & (b = c) → a = c (a = b) & (b = c) → a = c "transitivity of equality relation." For rules of inferences, if you have the rule called modes ponens in your system, it tells you that you can deduce the sentence αα from the sentences β → α, β.

Then you can construct deductions, those are lists of sentences. every sentence is either a logical axiom or non-logical axiom or a sentence that is deduced from some earlier sentences by means of a rule of inference. A sentence σ is provable if there is a deduction whose last sentence is σ.

For the semantic part, this is how you interpret the symbols, for example, you can say that your symbols represent natural numbers and that the symbol # represents addition of numbers and so on. Some sentences will be true under some interpretations and false under some others. Some will be true under ALL of the interpretations "those are called, valid sentences" and some will be false under any interpretation.

Now, completeness theorem says that, if you are given a sentence which is valid i.e. true under any interpretation, then you will find a deduction which ends up with the that sentence. What does that mean is the you will find a proof for every valid sentence.

Here is an illustration

> In order to illustrate Gödel's Completeness Theorem, I'll give an example. Suppose that we work in a language that has the symbols 0, 1, +, -, and *. In this language, we have the following axioms which I will collectively refer to as F:
>
> 1) $0 + a = a$
> 2) $a + (b + c) = (a + b) + c$
> 3) $a + (-a) = 0$
> 4) $a + b = b + a$
> 5) $1 * a = a$
> 6) $a * (b * c) = (a * b) * c$
> 7) for any a not equal to 0, there exists some b with $a * b = 1$
> 8) $a * b = b * a$
> 9) $a * (b + c) = (a * b) + (a * c)$
> 10) 0 does not equal 1
>
> If you have some familiarity with Abstract Algebra abstract algebra, then you might recognize these as the field axioms. Now there are many mathematical frameworks in which the above axioms are true. For example, if we are working in the rational (fractional) numbers Q, then all of the above statements are true (when we interpret 0, 1, +, -, and * in the usual way).
>
> Similarly, all of the above statements are true if we are working in the real numbers R or the complex numbers C. On the other hand, if we're working the integers Z, then statement 7) above is not true (there is no integer n such that $2 * n = 1$).

Logicians call a mathematical framework (or mathematical universe) that satisfy these axioms a *model* of the axioms. Hence, each of Q, R, and C are models of F, but Z is not a model of F.

Now one would hope that if we could prove a statement from the axioms F, then that statement should be true in any model of F. That is, our proof system is "sound" in the sense that if we can prove a statement from F, then that statement should logically follow from F. This fact is true and is called the Soundness Theorem. For example, one can prove the statement "If a + a = a, then a = 0"

From the above axioms F, and sure enough, this is true in each of Q, R, and C. The really interesting question is the converse, i.e. if a statement is true in every model of F, must it be the case that we can prove it from F?

For example, the statement S = "There exists an a with a*a = 2" is true in R, but false in Q since the square root of 2 is irrational. Similarly, the statement T = "There exists an a with a*a = -1" is true in C, but false in R (the imaginary unit i satisfies the statement in C). Hence, if you believe the Soundness Theorem, we should not expect to be able to prove either S or (not S) from F because there is one model of F in which S is true, and one where S is false. Similarly, we should not expect to be able to prove either T or (not T). Thus, our system F is incomplete, i.e, there are statements X such that we can neither prove X nor (not X). However, there are damn good reasons why it is incomplete because there are statements which can be either true or false depending on which model of F you are currently working.

> The Completeness Theorem basically says that this is the only way a system can be incomplete. In other words, the above converse question is true, which implies that we can prove absolutely everything that is not ruled out for the above basic reason.

As impressive as is the Completeness Theorem, this feigns in comparison with Gödel's Incompleteness Theorem, published 1931 with the title; "Über formal unentscheidbare Sätze der *Principia Mathematica* und verwandter Systeme" ("On Formally Undecidable Propositions of *Principia Mathematica* and Related Systems.")

A masterpiece, I think that ranks in scientific folklore with Einstein's relativity and Heisenberg's uncertainty. Promulgated in Vienna in the early nineteen-thirties, the notion of incompleteness threw mathematics into a hall of mirrors, where it reflected upon itself to alluring, if disorienting, effect: the theorem proved, using mathematics, that mathematics could not prove all of mathematics. Of course, it has a proper and technically precise formulation, but the late logician Verena Huber-Dyson paraphrased it for me as follows: "There is more to *truth* than can be caught by *proof*." Or, as the British novelist Zia Haider Rahman put it in his award-winning début, "In the Light of What We Know," "Within any given system, there are claims which are true but which cannot be proven to be true."

Roughly speaking, this theorem established the result that it is impossible to use the axiomatic method to construct a mathematical theory, in any branch of mathematics, that entails all of the truths in that branch of mathematics. It tells us that there are mathematical "blind spots": parts of mathematics that traditional methods of proof cannot access. These results are thought by many to have far-reaching consequences for computing and for our understanding of the nature of the human mind. Godel's results have thus been the subject of a great deal of popular attention. Indeed, few other results in the history of mathematics have had such an impact outside of mathematics.

In England, Alfred North Whitehead and Bertrand Russell had spent years on such a program, which they published as *Principia Mathematica* in three volumes in 1910, 1912, and 1913.) For instance, it is impossible to come up with an axiomatic mathematical theory that captures even all of the truths about the natural numbers (0, 1, 2, 3,.) This was an extremely important negative result, as before 1931 many mathematicians were trying to do precisely that - construct axiom systems that could be used to prove all mathematical truths. Indeed, several well-known logicians and mathematicians (e.g., Whitehead, Russell, Frege, Hilbert) spent significant portions of their careers on this project. Unfortunately for them, Gödel's theorem destroyed this entire axiomatic research program.

Godel's Incompleteness, in the wrong hands has been shown to prove and disprove the existence of God, the correctness of religion and its incorrectness against the correctness of science. The number of opportunistic arguments carried out in the name of Godel's incompleteness theorem is numerous to count.

The theorem in a nutshell

> If we have a list of axioms which we can enumerate with a computer, and these axioms are sufficient to develop the basic laws of arithmetic, then our list of axioms cannot be both consistent and complete. In other words, if our axioms are consistent then in every model of the axioms there is a statement which is true but not provable.

The reception of Gödel's results was mixed. Some important figures in the field of logic and the foundations of mathematics quite quickly assimilated the results and understood their relevance, but there was also quite a lot of misunderstanding and resistance.

Gödel revealed his results to Carnap in Vienna on 26 August 1930, and announced his result (the first theorem) in a casual discussion remark in the famous Königsberg Conference on September 7, 1930. John von Neumann, who was in the audience and was at the time working in the context of Hilbert's program, immediately understood the great importance of the result. On 20 November he wrote a letter to Gödel on a "remarkable" corollary of Gödel's result he had discovered: the unprovability of consistency (the second theorem.)

Meanwhile Gödel himself, however, had found the same idea and had already sent the final version of his article, which now contained also a statement of the second incompleteness theorem, for publication. As mentioned, the article was published in January 1931. The word quickly started to spread of these results which apparently had great importance for the foundations of mathematics, though views on what really was the moral varied. Paul Bernays, perhaps the most important collaborator of Hilbert, showed great interest in the results, though he first had difficulties in understanding them properly. His active correspondence with Gödel also shows that Gödel was already at the time fully aware of the undefinability of truth.

> About incompleteness, Russell admitted the same. He wondered, "Are we to think that 2 + 2 is not 4, but 4.001?"

Although Gödel published his results eighty-five years ago, the theorem endures in the popular imagination. In June 2016, a group of latter-day Gödelians convened on Thursday nights for a crash course in incompleteness, on the roster at the Brooklyn Institute for Social Research. One syllabus text promised "Gödel Without (Too Many) Tears," while supplementary reading unpacked his work "in Words of One Syllable." According to "Gödel's Theorem: An Incomplete Guide to Its Use and Abuse," the chameleon interpretations emerge in discussions of math, philosophy, computer science, and artificial intelligence, as one might expect, but also in ruminations on physics, evolution, religion, atheism, poetry, hip-hop, dating, politics, and even the Constitution. Famously, Gödel informed the judge at his U.S. citizenship hearing about an inconsistency that he had discovered in the Constitution, which would allow a dictator to rise in America.

The students in the incompleteness class, held at the Brooklyn Commons, included a computer scientist obsessed with recursion (that is, self-referential things, like Russian nesting dolls or Escher's drawing of a hand drawing a hand); a public-health nutritionist with a fondness for her "Philo_sloth_ical" T-shirt; a philosopher in the tradition of American pragmatism; an ad man versed in the classics; and a private-school teacher who'd spent a lonely, life-changing winter reading Douglas Hofstadter's "Gödel, Escher, Bach," which won the Pulitzer in 1980. Even Hofstadter, a professor of cognitive science at Indiana University, still has Gödel on the brain. He delivered two talks on the logician recently, billing him as "The True Inventor of Programming Languages and Data Structures."

That the sun was setting on this golden period of Gödel's, if not Vienna's, if not indeed Europe's existence, where it is mentioned that when Gödel obtained his Austrian citizenship in 1929, the year he turned in his thesis containing the completeness theorem the political situation in Vienna was becoming increasingly violent, with riots in 1927 leaving 89 people dead and the conservatives in essentially open war with the socialist majority.

By the end of the decade (1930s) both Gödel's advisor Hans Hahn and Moritz Schlick had died (the latter was assassinated by an ex-student), two events which led to a personal crisis for Gödel. Also, his appointment at the University, that of Privatdozentur (Adjunct professor) was cancelled, being replaced by the position "lecturer of a new order," granted to candidates only after they had passed a racial test.

However, while Gödel continued to tour and lecture, his mental health was becoming increasingly unstable. In 1938, he had barely returned to lecturing after suffering from a particularly bad depressive episode when Nazi Germany annexed Austria. Unable to secure a position at the University of Vienna and facing conscription, Gödel was actually found fit for military service by the Nazi government in 1939, he married his long-time girlfriend, a dancer Adele Nimbursky, and move with her to the United States. During this time, Gödel's three trips to the United States triggered an investigation.

Upon arrival Gödel took up an appointment as an ordinary member at the Institute for Advanced Study; he would become a permanent member of the Institute in 1946 and would be granted his professorship in 1953. He would remain at the Institute until his retirement in 1976 - the Gödels never returned to Europe.

Gödel's early years at the Institute were notable for his close friendship with his daily walking partner Albert Einstein, as well as for his turn to philosophy of mathematics, a field on which Gödel began to concentrate almost exclusively from about 1943. The initial period of his subsequent lifelong involvement with philosophy was a fruitful one. It is presumed in what was of the time a neglect by the analytically-oriented American philosophical establishment of Gödel's significant contributions to metaphysics and insofar as Einstein is presumed to share Gödel's "German Bias for Metaphysics."

In 1949 he published his "A Remark on the Relationship between Relativity Theory and Idealistic Philosophy." The latter paper coincided with results on rotating universes in relativity he had obtained in 1949, which were first published in an article entitled: "An Example of a New Type of Cosmological Solutions of Einstein's Field Equations of Gravitation." A science fiction writer's delight - this solution has many unusual properties, in particular, the existence of closed time-like curves, where time travel is as ordinary as visiting the corner store. For me, holding an image of a rotating universes is up there with picturing the 4th dimension.

For one thing, how does one test empirically the rotation of the universe as a whole, there is nothing else for it to be rotating relative to. But, a big but, General Relativity is not *Machian*, and it offers a variety of ways in which an observer inside a sealed box can detect whether it is rotating. And, there are alternative theories of gravity, such as Brans-Dicke gravity, that are more Machian than General Relativity, per say. In these theories there is probably no meaningful sense in which the universe could rotate.

Nevertheless, there are cosmologies in which the universe is rotating according to general relativity. And as such, one of the earliest solutions to the Einstein field equations is the Gödel metric, which, amongst other things, establishes a rotating the universe. A rotating universe does not have to have a centre of rotation, and it can be homogeneous. In other words, we could determine a direction in the sky and say that the universe was rotating counter-clockwise at a certain rate about the line connecting us to that point on the celestial sphere. However, aliens living across the other side of the universe could do the same thing. Their line would be parallel to ours, but there would be no way to tell whether one such line was the real centre of rotation.

Over the years there has been quiet deliberation over whether Einstein was already familiar with predictions field equations Gödel used or whether he was taken by surprise when presented with Gödel's results. Palle Yourgrau, in his

"A World Without Time, The Forgotten Legacy of Gödel and Einstein" for example, is not alone in propagating the myth that Einstein was taken by surprise when presented with Gödel's results. Stephen Hawking states: "It was therefore a great shock to Einstein when, in 1949, Kurt Gödel...discovered a solution that represented a universe full of rotating matter, with closed time-like curves through every point."

But in 1914, almost as soon as Einstein realized the need to introduce a non-flat, dynamical space-time metrical structure, and well before he arrived at the final form of his field equations, he worried in print about the problem of closed time-like world-lines. Here is what Einstein wrote then:

"I shall now raise an even deeper-reaching question of fundamental significance, which I am not able to answer. In the ordinary theory of relativity, every line that can describe the motion of a material point, i.e., every line consisting only of time-like elements, is necessarily non-closed, for such a line never contains elements for which dx 4 [space co-ordinate differences] vanishes.

An analogous statement cannot be claimed for the theory developed here. Therefore a priori a point motion is conceivable, for which the four-dimensional path of the point would be an almost a closed one. In this case one and the same material point could be present in an arbitrarily small space-time region in several seemingly mutually independent exemplars. This runs counter to my physical imagination most vividly. However, I am not able to demonstrate that the theory developed here excludes the occurrence of such paths."

4: The Grasshopper and the Bull

A meek Grasshopper settles down on the horn of a vainglorious Bull. After a while the Grasshopper decides to jump off the Bull. Before it does so, it asks the Bull, is it okay for me to leave? The Bull doesn't notice the Grasshopper, so replies "I didn't even know you had come, and I shall not miss you when you go.

Patients received daily and increasing doses of insulin, rising to many hundreds of units, for a 6-week period. The depth of the resulting hypoglycaemic coma was determined by the patient demonstrating a Babinski response over a period of 15 minutes. They were then revived by ingesting glucose. Although this treatment was reserved for people experiencing their first attack of Schizophrenia, that wasn't always the case. Some patients didn't regain consciousness, some died.

John Forbes Nash Jr. underwent insulin coma therapy.

According to Nash his delusional state of mind was like living a dream. "Well I knew where I was, I was there an observation but I was starting to think that I was like a victim of a conspiracy." He goes on, "delusions often took a cosmic quality, a feeling of ominousness." Things would happen around him and they took on a tremendous significance. In madness, he saw himself as some sort of a messenger of having special function. "The Muslim concept with Muhammad, the messenger of Allah." As an atheist notwithstanding, a man of "great religious importance, a foot of God."

In hospital, he was asked, how it was to be a mathematician, as someone committed to rationality, and yet accept the communications of supernatural aliens. To Nash, the idea of aliens could not be resolved, they came to him in the same way as did his mathematical ideas. But Nash struggled with finding connectedness his entire life. Adolescence wasn't easy for an intellectually precocious boy with few social skills or athletic interests to help him blend in normality, even nominally.

A childhood friend, Donald Reynolds, who lived across the street from the Nashes, in Bluefield, West Virginia remembers how as kids they took Nash's "experimenting" as a sign of craziness – he was "Big brains." Nash, however remembers things differently. In a 2002 interview with PBS, when he was in his 60's, having fought his daemons most of his adult life recalls being called "bug brains." How his ideas "were sort of buggy or not perfectly sound." Nash is looking out and back into his self through the haze and murky mind.

In spite of his illness, or because of it, Nash concentrates on the only thing he knows - mathematics and he's high-flown goals. A striving for success, for acceptance, and for recognition. Of his work, Nash: "I had some recognition. I was making some progress professionally, but I wasn't really at the top I didn't have top-level recognition." Without doubt he was bitterly disappointment when he learnt that he had, again, been overlooked for the 1958 Fields Medal. His work on parabolic and elliptic equations was still unpublished when the committee made their choice.

After months of bizarre behaviour, Alicia, his wife had Nash involuntarily hospitalised at McLean Hospital, a private psychiatric hospital outside of Boston in 1959. Less than two years after his release from McLean, Nash was hospitalized once again, this time at Trenton State Hospital, the former New Jersey Lunatic Asylum. When his colleagues heard where Nash was, many were outraged. "Who's going to figure out what is wrong with a genius there" asked one. "It is in the national interest," warned another, "that everything possible be done to protect Nash's exceptional mind." His mother, Virginia, visited Nash but could hardly bear to see her son in such a state.

Of course, they were right, but treating delusional disorders is extraordinarily difficult. In Nash, in a man [suffering] extended periods of paranoid schizophrenia; psychoeducation, psychotherapy, critical in treatment was not to be provided effectively. How could he commit to forging a trusting relationship when by in large he was subjected to a less than controversial aggressive insulin coma therapy which by 1961 had been phased out in all but a few hospitals.

Nash's psychotic process allowed a clinical refuge permitting his high-flown goal – to be the world's greatest mathematician, not just in this world, but in his world. Nash comments; "to some extent, sanity is a form of conformity" a distinction thus made between "consensual reality" and "private logic." He continues: "people are always selling the idea that people who have mental illness are suffering. But it's really not so simple, I think mental illness or madness can be an escape." The idea of escape reflects a function of an arrangement and protection. Not only did his delusions reflect his high-flown goal but his experience of auditory and visual hallucinations. He said: "you're really talking to yourself, is what the voices are." Interestingly, this explanation is essentially the same as the sub-vocalization hypothesis of auditory hallucinations.

Later, a lucid Nash would write "even when I was mentally disturbed, I had a lot of interest in numbers. I began to think more scientifically as to the years like the 80s, and maybe the later 70s. And so, there's a transition from really having more of an enthusiasm for the numbers, like maybe magical or representing a divine revelation, and just a more scientific appreciation of numbers, and these are not necessarily entirely far apart."

A cautious Nash draws a connection between great mathematicians and maniacal characteristics, delirium and symptoms of schizophrenia - between genius and madness. His hesitation, possibly a contemplation of his own place in the broader mathematics community. He thought of himself as superior, intellectually, mathematically superior. But how was he being viewed by others, those Nash considered intellectually equal? Had his illness affected his professional progress. At about the age of 30 a sternly competitive Nash was now pathologically anxious. By his own standards, Nash had fallen short. For a decade, he had pursued the Fields Medal, mathematics' highest honour; that year, he failed to win it again.

As we know core schizophrenia begins in adolescence or early adulthood, may involve a psychotic break, certainly involves diminished executive function, affective blunting and a thought disorder. Could Nash have been misdiagnosed. We're not completely certain.

But 1959 was probably the very worst time in the history of American psychiatry to become ill. Psychiatry then was still drenched in Freudian dogma. For the Freudian psychoanalysts' "Schizophrenia" was really a wastebasket diagnosis: It was used indiscriminately on all patients who did not seem to be suitable candidates for "the couch."

Virtually every patient apparently incapable of having a "transference relationship" was called "schizophrenic," and the inheritance of this ghastly tradition is still with us today. Transference means coming to see your therapist as your parent. Nash was certainly delusional and evidently experienced hallucination as well. He filled the blackboards of Fine Hall at Princeton with indecipherable scribblings, and wandered about the campus in an apparent daze. He became known as "the phantom of Fine Hall."

"His descent into madness had been sudden; his reawakening was gradual, almost imperceptible."

Sylvia Nasar
2002

John Nash began as an introverted boy who preferred to play in solitude in what was a town that had seen ebbs and flows depended almost entirely upon Norfolk and Western Railroad. When coal tonnage was good and the market for coal was booming, Bluefield was essentially a "little New York," as it was called in the day. A bustling metropolis, it had a nightlife and a personality that was "a little bit Chicago, a little bit New York, and a whole lot of Pittsburgh" rugged and with steel and coal embedded in its soul.

The coal boom generated a flood of money in the area. Nearby Bramwell, incorporated in 1888, boasted that it was the "Millionaires' town" because more millionaires per capita lived there than anywhere in the Nation. The city also had more automobiles per capita than any other city in the country.

During the 1920s, the twelve-story West Virginian Hotel was built. It has been adapted and in the 21st century is operated as the West Virginia Manor and Retirement Home. In 1924, nearby Graham, Virginia decided to rename itself as Bluefield to try to unite the two towns, which had been feuding since the civil war. The strategic importance of the city was so great that Adolf Hitler put Bluefield on his reputed list of German air raid targets in the United States.

Nash showed little interest in playing with other children favouring instead books and finding social events as tedious distractions from his books and experiments. Both his parents responded enthusiastically encouraging his education, both by seeing that he got good schooling and also by teaching him herself. Often Boredom and simmering adolescent aggression led him to play pranks, occasionally ones with a nasty edge.

Johnny's teachers at school certainly did not recognise his genius while his classmates noticed his mental health problems early. it would appear that he gave his teachers little reason to realise that he had extraordinary talents. They were more conscious of his lack of social skills and, because of this, labelled him as backward. By the time he was about twelve years old he preoccupied himself within doing scientific experiments. It was fairly clear that he learnt more at home than he did at school.

Nash first showed an interest in mathematics when he was about 14 years old. Quite how he came to read ET Bell's Men of Mathematics is unclear but certainly this book inspired him. He tried, and succeeded, in proving for himself results due to Fermat which Bell stated in his book.

Nash won a scholarship to the Carnegie Institute of Technology, to study chemical engineering. Soon, however, his professors recognised his remarkable mathematical talents and persuaded him to specialise in mathematics. He received a BA and an MA in Mathematics in 1948.

When he applied for graduate school, one of his professors, Richard Duffin, was less than superfluous when writing a recommendation; "This man is a genius." He was accepted to study for a doctorate at Harvard, Princeton, Chicago and Michigan; but when Princeton offered him the most prestigious fellowship it had, Nash made his decision to study there.

At Princeton, Nash avoided attending lectures and seldom read books, preferring not to learn mathematics according to him "second-hand". Instead, he worked things out from first principles while pacing endlessly, whistling Bach, or riding bicycles in tight concentric circles. In 1950, Nash received his doctorate from Princeton with his thesis, "Non-Cooperative Games."

Nash's 1950 paper acknowledges Von Neumann and Morgenstern's work; the theory of two-person zero-sum games of a type which Nash called "cooperative." That theory is based on an analysis of the interrelationships of the various coalitions which can be formed by players of the game. Nash's theory, in contradistinction, is based on the absence of coalitions. it is assumed that each participant acts independently, without collaboration or communication with any of the others. The notion of an equilibrium point (the Nash equilibrium) "is the basic ingredient in our theory," writes Nash.

The foundations for the theory were laid in a monumental study by John von Neumann and Oskar Morgenstern called Theory of Games and Economic Behaviour (1944.) Nash was the first to emphasise a distinction between cooperative games, as studied by von Neumann and Morgenstern (where the participants negotiate with one another) and non-cooperative games, where there is no such negotiation

Joel Watson provides an illustration of the equilibrium and its relation to security strategies (the context of Nash's work: von Neumann and Morgenstern.)

Fig 1. Players' potential strategy profiles

1\2	X	Y	Z
A	2,3	1,2	6,5
B	1,0	0,2	4,0
C	3,4	2,2	2,0

Consider the game depicted in Fig 1. In this game, C and Y are the players' security strategies, so a naive application of von Neumann and Morgenstern's maximin theory (absent binding agreements) would predict that strategy profile (C, Y) be played. However, this strategy profile is plainly inconsistent with the idea that players are rational in responding to each other. In particular, if player 1 is expected to select C then player 2 behaves quite irrationally by choosing Y. In fact, strategy Y is not even rationalizable for player 2; it does not survive iterated removal of dominated strategies (see below.) Thus, the notion of a security strategy is not a good theory of behaviour for non-zero–sum games, demonstrating the limits of von Neumann and Morgenstern's analysis.

Next, observe that the game has two Nash equilibria in pure strategies, (C, X) and (A, Z). Both of these are reasonable predictions in the sense that, in both cases, the players are best responding to one another. For example, if player 1 is sure that player 2 will select X, then it is best for player 1 to select C; likewise, if player 2 is convinced that player 1 will select C, then it is optimal for player 2 to choose X. There is also a mixed-strategy Nash equilibrium in which player 1 randomizes between A and C, and player 2 randomizes between X and Z. That the game has multiple

Nash equilibria demonstrates the general economic problem of coordination, in particular the possibility that the players will coordinate on the less efficient Nash equilibrium. Other games, such as the Prisoner's Dilemma, have only inefficient equilibria and thus reveal a fundamental tension between individual and joint incentives.

One of Game theory's best-known example is the prisoner's dilemma (see Fig 2.) Two criminals in separate prison cells face the same offer from the public prosecutor. If they both confess to a bloody murder, they each face ten years in jail. If one stays quiet while the other confesses, then the snitch will get to go free, while the other will face a lifetime in jail. And if both hold their tongue, then they each face a minor charge, and only a year in the clink. Collectively, it would be best for both to keep quiet. But given the set-up, an economist armed with the concept of the Nash equilibrium would predict the opposite: the only stable outcome is for both to confess.

In a Nash equilibrium, every person in a group makes the best decision for herself, based on what she thinks the others will do. And no-one can do better by changing strategy: every member of the group is doing as well as they possibly can. In the case of the prisoners' dilemma, keeping quiet is never a good idea, whatever the other mobster chooses. Since one suspect might have spilled the beans, snitching avoids a lifetime in jail for the other. And if the other does keep quiet, then confessing sets him free.

Applied to the real world, economists use the Nash equilibrium to predict how companies will respond to their competitors' prices. Two large companies setting pricing strategies to compete against each other will probably squeeze customers harder than they could if they each faced thousands of competitors.

The Nash equilibrium helps economists understand how decisions that are good for the individual can be terrible for the group. This tragedy of the commons explains why we overfish the seas, and why we emit too much carbon into the atmosphere. Everyone would be better off if only we could agree to show some restraint. But given what everyone else is doing, fishing or gas-guzzling makes individual sense.

Fig 2. Illustration of the prisoner's dilemma

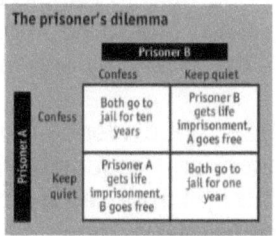

Beginning in the 1960s, economists began to successfully apply game theory to real-life situations. Strikes, collective bargaining, these situations of conflict and cooperation are part of the backbone of practical economics. Auctions, farm subsidies, monetary policy, international trade all were now seen as strategic games. By the late 1970s, game theory had become one of the foundations of modern economics. And at the centre was the Nash equilibrium. According to Sylvia Nasar, there are not more than ten ideas, in the post-war period, which you could say are equivalent. It had a huge impact in economics. It made economics a much more useful subject.

> "I knew it was good work, but you cannot know how much something will be appreciated in the future. You don't have that crystal ball."
>
> John Nash

By the 1980s, economists expected that game theory would be recognized with the Nobel Prize. Year after year, it didn't happen. The committee in Stockholm could not conceivably dream of giving a Nobel Prize if they couldn't include John Nash as one of the deserving people. Members of the Nobel committee worried that Nash was unstable and wouldn't be able to handle the pressures of the ceremony. Some even feared he might do something that would embarrass the Academy and tarnish the prize.

Although there may be scant evidence to suggest any firm links between brilliance and madness, Nash finds that some "rationality of thought imposes a limit on a person's concept of his relation to the cosmos", suggesting that his return to a normal mental state was somehow detrimental to his work as a mathematician.

However, university lecturer in experimental Psychology, Dr Jennifer Lau gives a sobering verdict on the possible connection between genius and madness that these works play upon. "There seems to be a causal link between genius and the subsequent mental illness (being assumed), and I don't think there really is that established link." Rather than one being a result of the other, these traits sometimes "co-occur in the same person". Whilst one can be genetically predisposed to getting a mental illness, there is no genetic reason that mathematical ability should be yoked to, say, psychosis or schizophrenia.

What is psychologically plausible in Proof is the daughter Catherine's descent into mental illness, which is triggered by her fear that she will go mad like her father. Lau highlights the danger of "looking out for signs of the disorder in yourself," which could constitute a "modelling" of behaviour. "It may be the case that you become what you think you are," she says, gravely.

Nash, "I would not dare to say that there is a direct relation between mathematics and madness," he goes on; "but there is no doubt that great mathematicians suffer from maniacal characteristics, delirium and symptoms of schizophrenia." Nash is one of the high-profile mathematicians whose life-story, portrayed in the film A Beautiful Mind, contributes to the popularly held belief that there is a link between genius and madness. The same high-profile figures from the discipline of mathematics, such as himself, "seem to have made a disproportionate contribution to this popular assumption." One thinks of Isaac Newton sticking a needle into the back of his eye (he wanted to find out how this would distort his vision) or more recently, the eccentricities of Grigori Perelman.

As well as A Beautiful Mind (1998) the 2014 film, Imitation Game was based on the biography Alan Turing: The Enigma by Andrew Hodges. Madness and genius populate a particular genre - films perhaps exaggerating eccentricity. Darwin's many lifelong and serious illnesses have been the subject of much speculation and study for over a century. Darkest Hour, based on Winston Churchill's account of his early days as Prime Minister during World War II hints at Churchill's ability to overcame his manic-depressive illness.

Nash goes on, "but after my return to the dream-like delusional hypotheses in the later 60s I became a person of delusionally influenced thinking but of relatively moderate behaviour and thus tended to avoid hospitalization and the direct attention of psychiatrists."

Gradually, he begun to intellectually reject some of the delusionally influenced lines of thinking which had been characteristic of my orientation. This began, most recognizably, with the rejection of politically oriented thinking as essentially a hopeless waste of intellectual effort. So, "at the present time, 1995, I seem to be thinking rationally again in the style that is characteristic of scientists) albeit, finding that he felt more limited, presumably referring to his mathematics.

In rejecting, renouncing his delusional hypotheses and reverting to thinking of himself as a human of more conventional circumstances – interludes of, as it were, enforced rationality, Nash did succeed in doing, according to him "some respectable mathematical research."

The concomitant consequences of such a commonly held viewpoint mentioned above are manifold. For one, even highly knowledgeable people, who may have all the facts with them, may respond in irrational ways should they, at some point in their lives, encounter any mental illness. This is a sad but a true fact. The problem of mental illness is further compounded by the fact that both undergraduate and graduate students, who may suffer from various forms of mental illness to varying degrees, do not usually have access to the necessary tools to combating the same. And, even if they do, there might be considerable reluctance on their part to using those tools for varied reasons.

Now, this does not really happen in core adolescent-onset schizophrenia. Some of the patients don't recover at all; others make only a social recovery, ending with what the Europeans call a "defect," waking magically from classical schizophrenia and go on to have a normal life. Could Nash have been given the wrong diagnosis? Or rather that his psychoanalytically oriented clinicians in 1959 gave the wrong diagnosis and ever since this has been unthinkingly accepted.

This is the way psychiatry often works. The field has trouble with new ideas, unless they are heavily promoted by the pharmaceutical industry (think "neurotransmitters.") In 1893 Emil Kraepelin in Heidelberg popularized the concept of psychosis of adolescent onset as "dementia praecox," premature dementia, the premature part meaning adolescence or young adulthood. Then Eugen Bleuler, professor of psychiatry in Zurich, re-labelled Kraepelin's dementia praecox as "schizophrenia" in 1908, and detached it from age.

We still have Bleuler's "schizophrenia" with us today, more than a hundred years later. The field has made virtually no progress in unpacking chronic severe illness and differentiating out several distinct entities. In no other field of medicine would this be conceivable! DSM-5, the current edition, still refers to "schizophrenia" in the singular.

To be sure, other efforts at unpacking have been made, but they haven't caught on. In 1957 East German psychiatrist Karl Leonhard proposed a complicated alternative to the Kraepelin-Bleulerian standard that still has some acolytes today. In the Leonhardian scheme, Nash would probably have received the diagnosis "affective paraphrenia," but Leonhard said they don't recover, and Nash did. Someday his patient records will be available for scholarly analysis, and then we'll know a lot more.

Say, dost thou mourn thy ravish'd mate, that oft enamour'd on thy strains has hung? Or has the cruel hand of fate bereft thee of thy darling young?

An evening address to a nightingale
Cuthbert Shaw, 1809

On May 23rd 2015, four days after receiving the Abel Prize, on the way home from Norway, John Nash, along with his wife, was killed in a taxi crash on the New Jersey Turnpike. The taxi driver who lost control of his vehicle had only been driving taxis for two weeks. The American-Canadian mathematician, Louis Nirenberg, who was also that year's Abel Prize Laureate said he'd chatted with the couple for an hour at the airport in Newark before they'd gotten a taxi.

The Nashes were supposed to have been met by a limo at Newark Airport on Saturday but hailed a yellow taxi back to Princeton after they switched their flight at the last minute and arrived in New Jersey five hours early, five hours before their tragic death.

The Massachusetts Institute of Technology (MIT) club of Princeton had planned a celebration of the Abel Prize award Wednesday night, but instead conducted a tribute to the Nashes. While RAND Cooperation's president and CEO Michael Rich spoke of the sadness over the news of Nash's death. He went on "At a time when RAND, along with much of America, was grappling with the strategic implications of the new Cold War, John contributed important insights into game theory, which widened and deepened our understanding of mutual deterrence and the nuclear arms race." Michael Rich, with civility and respect talked not of Nash's arrest for indecent exposure. The charges were later dropped, but Nash was stripped of his top-secret security clearance and consequently fired from RAND Corporation.

Princeton University President Christopher Eisgruber said the Nashes were special members of the university community. "John's remarkable achievements inspired generations of mathematicians, economists and scientists who were influenced by his brilliant, ground-breaking work in game theory, and the story of his life with Alicia moved millions of readers and moviegoers who marvelled at their courage in the face of daunting challenges."

In September 1949, the world learned that the Soviet Union had joined the United States as a nuclear power. The shocking news intensified fears in the U.S. and put a premium on mathematicians. Mathematicians had helped win WWII; now there was hope they could protect America's strategic edge. Princeton University boasted the most elite math department in the world; each of its graduate students was handpicked. That year, one stood out: a twenty-year-old from West Virginia named John Forbes Nash.

According to RAND Corporation, it's a research organization that develops solutions to public policy challenges to help make communities throughout the world safer and more secure, healthier and more prosperous. That may well be so, but by the 1960s, America's rivals were paying attention. The Soviet newspaper *Pravda* nicknamed RAND "the academy of science and death and destruction." American outfits saw RAND as *a shadowy think tank in Santa Monica, California and* preferred to call them the wizards of Armageddon.

In recruitment ads, RAND bragged about its intellectual genealogy, tracing a direct line from its president, Frank Collbohm, to Isaac Newton. Whether or not that claim was true, the institute secured a reputation as the place to dream up new ways to wage and enemies at bay. Breakthroughs conceived at RAND the Internet, the first satellite to orbit Earth, and the modern computer.

In the summer of that year Nash worked for the RAND Corporation, an ultra-secret nuclear think-tank in Santa Monica, California, where his work on game theory made him a leading expert on Cold War military and diplomatic strategy, which dominated RAND's work.

During his time at MIT Nash's personal life became complicated. He met Eleanor Stier and they had a son, born in June 1953. Nash did not want to marry Eleanor although she tried hard to persuade him. Then, in the summer of 1954, while working for RAND, Nash was arrested in a police operation to trap homosexuals. He was dismissed from RAND.

For Nash to receive one of his field's highest honour only days before his death marked a final turn of the cycle of astounding achievement and jarring tragedy that seemed to characterize his life. At heart, however, and according to David Gabai, the Hughes-Rogers Professor of Mathematics and department chair at Princeton, Nash was a devoted mathematician whose ability to see old problems from a new perspective resulted in some of his most astounding and influential work. Gabai says. "It went beyond proving great results. He had a profound originality as if he somehow had insights into developing problems that no one had even thought about."

According to Camillo De Lellis at the Institute of Mathematics University of Zuerich, Nash had written very few papers. And, in fact Nash's fundamental contributions can be stated in very few lines. After his famous PhD thesis in game theory, Nash directed his attention to geometry and specifically to the classical problem of embedding smooth manifolds in the Euclidean space. Leaving aside Nash's celebrated PhD thesis the following four fundamental papers have started an equal number of revolutions in their respective topics.

The 1952 note on real algebraic varieties, the 1954 paper on C^1 isometric embeddings, the 1956 subsequent work on smooth isometric embeddings and finally the 1958 Hölder continuity theorem for solutions to linear (uniformly) parabolic partial differential equations with bounded non-constant coefficients. It should be noted that people were not even aware there that interesting non-smooth embedding existed.

To try and understand isometric embeddings we need to go back to the work of Riemannian and probably a little before then. In 1828, Gauss proved his Theorema Egregium (remarkable theorem in Latin,) establishing an important property of surfaces. Informally, the theorem says that the curvature of a surface can be determined entirely by measuring distances along paths on the surface. That is, curvature does not depend on how the surface might be embedded in 3-dimensional space - curvature was an intrinsic property of a surface, independent of its isometric embedding in Euclidean space. This was extended to higher-dimensional spaces by Riemann (Gauss' student) and led to what is known today as Riemannian geometry. Riemannian set up the axiom to describe non-Euclidean geometry, the concept of Riemannian manifolds which is a geometry in which each point is in a neighbourhood resembling a distorted Euclidean space.

But ever since Riemannian's work a question has existed of whether Riemannian's abstract geometry is more general than embedded geometry. For instance, can one embed a hyperbolic space into Riemannian's geometry? We know that we can make approximations. We can have some partial geometries with some so-called singularities, like crests, like pseudosphere, and hyperbolic crochet (albeit a small representation.) We know embedded geometry is restricted. Blanusa proved that hyperbolic geometry, infinitely large can be imbedded in 6-space.

Consider a smooth closed manifold Σ of dimension n (where with closed we mean, as usual, that Σ is compact and has no boundary). A famous theorem of Whitney shows that Σ can be embedded smoothly in R^{2n}, namely that there exists a smooth map w: $\Sigma \to R^{2n}$ whose differential has full rank at every point (i.e., w is an immersion) and which is injective (implying therefore that w is a homeomorphism of Σ with w(Σ)).

To succeed in cracking open such problems mathematics takes a degree of obsession. To quote the mathematician Andrew Wiles: "Pure mathematicians just love to try unsolved problems - they love a challenge." It was Wiles who finally solved Fermat's Last Theorem, an infamous number conundrum. The young Wiles discovered Fermat in his local Cambridge library when he was a child. "Here was a problem, that I, a 10-year-old, could understand, and I knew from that moment that I would never let it go. I had to solve it." 'Not letting it go' is a sign of the single-minded devotion needed to pursue such apparently impossible dreams. From films to plays, art has courted this romantic notion of the mad genius.

Princeton University boasted the most elite math department in the world; each of its graduate students was handpicked. That year (1948) one stood out: a twenty-year-old from West Virginia named John Forbes Nash. These young mathematicians were all pretty cocky, but he towered over them in arrogance and confidence and also in eccentricity.

A colleague, Mel Hausner remembers Nash always an entity unto himself. When Nash walked into the room you knew that Nash walked into the room. I think he thought of himself as superior, intellectually, mathematically superior. We thought highly of ourselves and each other, but with Nash it was double. Nash was just very clearly above it. Nash was also extremely annoying.

Warren Ambrose writes in 1953

> *There's no significant news from here, as always. Martin is appointing John Nash to an Assistant Professorship (not the Nash at Illinois, the one out of Princeton) and I'm pretty annoyed at that. Nash is a childish bright guy who wants to be "basically original," which I suppose is fine for those who have some basic originality in them.*
>
> *He also makes a damned fool of himself in various ways contrary to this philosophy. He recently heard of the unsolved problem about imbedding a Riemannian manifold isometrically in Euclidean space, felt that this was his sort of thing, provided the problem were sufficiently worthwhile to justify his efforts; so he proceeded to write to everyone in the math society to check on that, was told that it probably was, and proceeded to announce that he had solved it, modulo details, and told Mackey he would like to talk about it at the Harvard colloquium.*
>
> *Meanwhile he went to Levinson to inquire about a differential equation that intervened and Levinson says it is a system of partial differential equations and if he could only [get] to the essentially simpler analogue of a single ordinary differential equation it would be a damned good paper - and Nash had only the vaguest notions about the whole thing.*
>
> *So, it is generally conceded he is getting nowhere and making an even bigger ass of himself than he has been previously supposed by those with less insight than myself. But we've got him and saved ourselves the possibility of having gotten a real*

> *mathematician. He's a bright guy but conceited as Hell, childish as Wiener, hasty as X, obstreperous as Y, for arbitrary X and Y.*

Ambrose, the author of this letter, and Nash had rubbed each other (Halmos) the wrong way for a while. They had played silly pranks on each other and Ambrose seems not to have been able to ignore Nash's digs in the way others had learned to do.

> *And, if you're so good, why don't you solve the embedding theorem for manifolds.*

From 1952 Nash had taught at the Massachusetts Institute of Technology but his teaching was unusual and his examining methods were highly unorthodox. His research on the theory of real algebraic varieties, Riemannian geometry, parabolic and elliptic equations was, however, extremely deep and significant in the development of all these topics.

> Nash did solve the problem.

And not only did he provide one proof – he gave 2 amazing proofs dealing with Non-smooth and smooth imbedding. People were not even aware that there was an interesting non-smooth embedding construct. Nash's proof of the isometric embedding problem came as a complete surprise to much of the mathematical community. His methods were revolutionary. The great mathematician Mikhail Gromov said that Nash's work on the embedding problem struck him to be "as convincing as lifting oneself by the hair". But after great effort, Gromov finally understood Nash's proof: at the end of Nash's lengthy argument, Gromov said, Nash "miraculously, did lift you in the air by the hair."

Two years after his counterintuitive C^1 theorem, Nash addressed and solved the general problem of the existence of smooth isometric embeddings in his other celebrated work. As in the previously we consider Riemannian manifolds (Σ, g), but this time of class C^k with $k \in \mathbb{N} \cup \{\infty\} \setminus \{0\}$: this means that there is a C^∞ atlas for Σ and that, in any chart the coefficients g_{ij} of the metric tensor in the local coordinates are C^k functions.

Nash's smooth isometric embedding theorem - Let $k \geq 3$, $n \geq 1$ and $N = n(3n+11)$. If (Σ,g) is a closed C^k Riemannian manifold of dimension n, then there is a C isometric embedding u:$\Sigma \to R$

Nash covered also the case of non-closed manifolds as a simple corollary of Theorem above but with a much weaker bound on the codimension. More precisely he claimed the existence of isometric embeddings for $N' = (n + 1) N$. His proof contains however a minor error as pointed out by Robert Solovay, albeit can be easily fixed using the same ideas, but at the price of increasing slightly the dimension N'.

Nash, as well as Louis Nirenberg were awarded the 2015 Able prize by the King of Norway for "For striking and seminal contributions to the theory of nonlinear partial differential equations and its applications to geometric analysis."

Nash began to realize that he wouldn't be getting out of the McLean unless he conformed and behaved "normally." So, he in part would do that as if he would be sweeping the delusions under a rug and they were able to come out later on, and could be triggered, and he would move very quickly to accepting it again. Nash retained a lawyer, who secured his release after fifty days of hospitalization. Within weeks, he resigned from MIT, withdrew all the money from his pension fund and announced he was leaving for Europe.

In July 1959, the Nashes arrived in Paris to find the city in turmoil. The streets reverberated with strikes, explosions and mass demonstrations against the nuclear arms race. A week later, Nash suddenly took off on his own. He went to Luxembourg and announced he wanted to give up his American citizenship. He was turned away. When Nash and his wife, Alicia, got to Geneva he thought of a way of being a refugee. Nash explains they had a slogan, "city of refuge." I envisioned a hidden world where the Communists and the anti-Communists were really the same, they were sort of schemers. I had the idea that some of the people, like Eisenhower and the Pope and the powers that be might be unsympathetic to me. These thoughts on the surface are not rational, but there could be a situation where there were - things were not what they might seem.

Most of the time when he was trying to give up his citizenship, he was being followed around by the naval attaché, who had as his commission to get his passport back and give it to Alicia. And so, he chased him around Europe. On returning to the US, the Nashes moved to Princeton. To support her family, Alicia took a job with a research division of RCA. She hoped that with the help of the math community, they could start over again.

In the year 1960, Nash's madness blew way out of proportion when right in the middle of his lecture he turned towards a very bright student and asked him to take over his class. Stating this, Nash went straight away to Washington DC. Two weeks later, a bewildered Nash arrived at MIT holding a New York Times edition, and announced that aliens were trying to contact him through coded messages encrypted in the paper, which only he could decipher. Alicia knew something was wrong with her husband but never imagined the magnitude of Nash's illness.

Unaware of Nash's delusions the student took it sportingly, but the problem was greater than it appeared to be. Day by day, John Nash's mental condition depleted. The McCarty episode often gave him nightmares that communists were sprawling through the country, targeting honest civilians to join their evil cause. Nash concluded that people wearing red neckties were members of "Crypto Communist Party". In his own words, Nash later described "The staff at my university, the Massachusetts Institute of Technology, and later all of Boston were behaving strangely towards me.... I started to see crypto-communists everywhere... I started to think I was a man of great religious importance, and to hear voices all the time. I began to hear something like telephone calls in my head, from people opposed to my ideas."

When incognizant of Nash's mental problems, University of Chicago offered him the job of a full-time professorship. Nash promptly wrote back that he denies taking up the position as he is scheduled to become "Emperor of Antarctica."

His behaviour at the university was now intolerable; he would attend his colleague's seminar and declare that his face was morphed by the face of Pope XXIII on the cover of life magazine. Under the dreadful clutches of paranoid schizophrenia, John Nash once made a complete fool of himself while conducting a seminar; in the middle of his presentation he started talking gibberish and walked out of the auditorium. His eccentricities now horrified Alicia when she found that Nash was writing strange letters to United Nation in blue, black, red and green ink advocating forming a "One World Government".

5: The Wolf and the crane

A wicked Wolf has a bone stuck in his throat. He persuades a Crane to put her head in his throat and remove the bone. When the Crane finally removes the bone, she demands a reward. The Wolf smiles and replies, surely you have been given enough reward by me not eating you.

It took the world's leading mathematicians several years to verify that Grigori Perelman had definitively solved the problem. A problem involving the deep structure of three-dimensional shapes and carrying a $1Million reward offered by the Clay Mathematics Institute. A problem that had eluded mathematicians for 100 years. But in 2006, the reclusive Russian abruptly quit the world of mathematics in disgust – declining to accept the coveted Fields Medal and then in June 2010, snubbed a ceremony in his honour at the Institut Océanographique, Paris to accept his $1 million prize.

He was quoted to say, "I'm not interested in money or fame; don't want to be on display like an animal in a zoo." "He has rather strange moral principles. He feels tiny improper things very strongly," says Sergei Kisliakov, director of St Petersburg's Steklov Mathematics Institute, where the maths prodigy once worked as a researcher. The handful of neighbours who have seen him paint a picture of a scruffily dressed, unworldly eccentric. "He always wears the same tatty coat and trousers. He never cuts his nails or beard. When he walks he simply stares at the ground, rather than looking from side to side," one told a Moscow newspaper. Perelman refuses to talk to journalists camped outside his home. When one managed to reach him on his mobile, he was told: "You are disturbing me. I am picking mushrooms."

An example of those tiny improper things Kisliakov refers to came to light in 2011. Perelman was quoted to say he was disappointed with the ethical standards of the field of mathematics. A 2006 New Yorker article implied that Perelman was referring in particular to the efforts of Fields medallist, Shing-Tung Yau to downplay his role in solving the problem and play up the work of Huai-Dong Cao and Xi-Ping Zhu.

In 2006, Yau, a professor of mathematics at Harvard and the director of mathematics institutes in Beijing and Hong Kong lectures at the Friendship Hotel to a gathering of several hundred physicists, including a Nobel laureate, assembled in an auditorium. Yau, a stocky man of fifty-seven, stood at a lectern in shirtsleeves and black-rimmed glasses and, with his hands in his pockets, described how two of his students, Zhu and Cao, had completed a proof of the problem only a few weeks earlier.

The interest by physicists to listen to a maths professor was clear, a non-trivial relationship exists between the Poincare' conjecture, the problem, and Einstein's general theory of relativity. As well, a key mathematical tool crucial in Perelman's proof, the topological method of Ricci flow was originally developed in Physics as the approximate renormalization group flow of the nonlinear 2d sigma model, and so directly related to string physics.

Yau would report, Zhu and Cao were indebted to his long-time American collaborator Richard Hamilton, who deserved most of the credit for solving the Poincaré conjecture. Hamilton's mathematical contributions are many, but he is best known for having created the Ricci flow method in 1981. Yau, Hamilton and Rick Schoen were invited to visit the Mathematical Sciences Research Institute (MSRI), Berkeley, CA the first year it opened. Yau did make some mention of Perelman, who, he acknowledged, had made an important contribution. Nevertheless, Yau said, "in Perelman's work, spectacular as it is, many key ideas of the proofs are sketched or outlined, and complete details are often missing."

Perelman had indeed learned much from Hamilton. In 1993, he began a two-year fellowship at Berkeley. While he was there, Hamilton gave several talks on campus, and in one he mentioned that he was working on Poincaré. After one of his talks at Berkeley, he told Perelman about his biggest obstacle - as a space is smoothed under the Ricci flow, some regions deform into what mathematicians refer to as "singularities." Some regions, called "necks," become attenuated areas of infinite density. More troubling to Hamilton was a kind of singularity he called the "cigar" which I will visit later in this chapter.

In 1995, Hamilton published a paper in which he discussed a few of his ideas for completing a proof of the Poincaré. Reading the paper, Perelman realized that Hamilton had made no progress on overcoming his previously stated obstacles - the necks and the cigars. "I hadn't seen any evidence of progress after early 1992," says Perelman. "Maybe he got stuck even earlier." However, Perelman thought he saw a way around the impasse. In 1996, he wrote Hamilton a long letter outlining his notion, in the hope of collaborating. "He did not answer," Perelman said. "So, I decided to work alone."

In the spring of 2003, after Perelman completed his lecture tour in the United States, Yau had recruited Zhu and another student, Cao, a professor at Lehigh University, to undertake an explication of Perelman's proof. Zhu and Cao had studied the Ricci flow method under Yau, who considered Zhu, in particular, to be a mathematician of exceptional promise. "We have to figure out whether Perelman's paper holds together," Yau tells them. Yau arranged for Zhu to spend the 2005-06 academic year at Harvard, where he gave a seminar on Perelman's proof and continued to work on his paper with Cao.

That year, the thirty-one mathematicians on the editorial board of the *Asian Journal of Mathematics* (AJM) received a brief e-mail from Yau and the journal's co-editor informing them that they had three days to comment on a paper by Zhu and Cao titled "The Hamilton-Perelman Theory of Ricci Flow: The Poincaré and Geometrization Conjectures," which Yau planned to publish in the journal. The paper had been accepted by the *A.J.M.*, and an abstract was posted on the journal's Website. But within a week, speculatively, Zhu and Cao's paper now had a new title; a very much grander, a loftier title - "A Complete Proof of the Poincaré and Geometrization Conjectures: Application of the Hamilton-Perelman Theory of the Ricci Flow." And, that's not all. The abstract had also been revised. A new sentence explained, "This proof should be considered as the crowning achievement of the Hamilton-Perelman theory of Ricci flow."

Yau and Yang Le, even thought it would be a good idea to quantify Zhu and Cao's contribution, bizarrely. 50% goes to Hamilton, 25% Perelman, and 30% Zhu and Cao. The numbers don't yet add up to 100% - the wrong precise values are probably a contribution of a journalistic blunder, to be fixed for sure in the fullness of time. When Perelman was asked whether he had read Cao and Zhu's paper, he had. "It is not clear to me what new contribution did they make," he said. "Apparently, Zhu did not quite understand the argument and reworked it."

Perelman added, "as for Yau, I can't say I'm outraged. Other people do worse. Of course, there are many mathematicians who are more or less honest. But almost all of them are conformists. They are more or less honest, but they tolerate those who are not honest." He has also said that "It is not people who break ethical standards who are regarded as aliens. It is people like me who are isolated."

Yau, Cao and Zhu's concern was to check Perelman's proof rather than race to scoop each other on it. The proper thing was to check the logic and point out gaps, give Perelman the time and opportunity to respond. Instead, it seems that Yau, who was working under the guise of a referee, was actually just trying to insert himself on the list of credits by filling the gaps himself and claiming ownership.

Whether Yau's contributions in "filling the gaps" are considered novel, worthy of their own merit, or whether his group had just performed the perfunctory calculations that Perelman only alludes to in his outline are widely disputed. Of course, when Yau announced that Perelman's key ideas of the proofs were sketched or outlined, he was right. Perelman's 2002 and 2003 papers are also compounded by the highly compressed manner in which he presented them.

Generally speaking, to say that a mathematician "filled gaps" to receive credit for original work is not all that straight forward. There are two ways to get credit for an original contribution. The first is to produce an original proof, clearly. The second is to identify a *significant* gap in someone else's proof and supply the missing chunk. However, only true mathematical gaps, missing or mistaken arguments can be the basis for a claim of originality. Filling in gaps in exposition - shortcuts and abbreviations used to make a proof more efficient does not (should not) count. When, in 1993, Andrew Wiles revealed that a gap had been found in his proof of Fermat's last theorem for example, the problem became fair game for anyone, until, the following year, Wiles fixed the error. Most mathematicians would agree that, by contrast, if a proof's implicit steps can be made explicit by an expert, then the gap is merely one of exposition, and the proof should be considered complete and correct.

The 2006, New Yorker article, Manifold Destiny, A legendary problem and the battle over who solved it described this period as a clash over the Poincaré conjecture. The issue most certainly caused a bitter controversy in the geometric analysis community and the article undoubtedly exacerbated that divide. Richard Hamilton, Louis Nirenberg, John Coates and various others wrote passionate letters defending Yau after the article was published because they thought that it depicted him unfairly.

On the other hand, Bruce Kleiner and John Lott (the two who wrote the best exposition of Perelman's proof) were furious about the authorship situation and at one-point Lott refused to share a stage with Zhu. Furthermore, Joan Birman wrote a letter saying that the episode involving the Asian Journal of Mathematics had left a "very public and very bad black mark" on mathematics.

As mentioned in the introduction, the controversy caused Perelman to leave mathematics entirely, now living in seclusion, away from the media in St Petersburg. Perelman is regarded with admiration for his integrity and grit. The fact that he didn't make any serious mistakes means that the authorship controversy is presumably completely settled. However, this wasn't clear when the papers were initially published. Extremely terse, the arguments incredibly technical, incredibly difficult.

In May 2006, a committee of nine mathematicians voted to award Perelman a Fields Medal for his work on the Poincaré conjecture. However, Perelman declined to accept the prize. Sir John Ball, president of the International Mathematical Union, approached Perelman in Saint Petersburg in June 2006 to persuade him to accept the prize. After ten hours of attempted persuasion over two days, Ball gave up.

In March 2010, Perelman was awarded a Millennium Prize for solving the problem. In June 2010, he did not attend a ceremony in his honour at the Institut Océanographique, Paris to accept his $1 million prize. He considered the decision of the Clay Institute unfair for not sharing the prize with Hamilton. "The main reason is my disagreement with the organized mathematical community. I don't like their decisions, I consider them unjust." According to those who have known Perelman since he was a teenager portray him as "impeccably honest." He believed in the absolute righteousness of mathematicians about giving credit where credit is due.

In December 2006, Cao and Zhu re-submitted their Geometrization Conjecture paper with a revised, a less haughty title, and a more considered abstract on arXiv.org, a repository of electronic preprints approved for publication after moderation. The new title: "Hamilton-Perelman's Proof of the Poincaré Conjecture and the Geometrization Conjecture." The revised abstract:

> "In this paper, we provide an essentially self-contained and detailed account of the fundamental works of Hamilton and the recent breakthrough of Perelman on the Ricci flow and their application to the geometrization of three-manifolds. In particular, we give a detailed exposition of a complete proof of the Poincaré conjecture due to Hamilton and Perelman."

Now it's at about this time that we start looking at the ceiling and saying "what was Poincaré's Conjecture again?" Come on, you remember - if in a closed three-dimensional space, any closed curves can shrink to a point continuously, this space can be deformed to a sphere.

If we stretch a rubber band around the surface of an apple, then we can shrink it down to a point by moving it slowly, without tearing it and without allowing it to leave the surface. On the other hand, if we imagine that the same rubber band has somehow been stretched in the appropriate direction around a doughnut, then there is no way of shrinking it to a point without breaking either the rubber band or the doughnut. The surface of the apple is said to be "simply connected," but that the surface of the doughnut is not. Poincaré, almost a hundred years ago, knew that a two-dimensional sphere is essentially characterized by this property of simple connectivity, and asked the corresponding question for the 3-dimensional sphere in 1904.

Although usually formulated as a pure mathematics problem about the classification of abstract manifolds, the Poincare conjecture is really about the structure of the space we live in, and the degree to which we humans, as prisoners within that space, unable to break free and observe it from the outside, may nevertheless determine some key features of its structure.

Poincare, himself, proposed a procedure he believed would enable us to make such a determination. It is not a practical procedure, and was never intended to be viewed as such. Thanks to Perelman, we now know that Poincare's proposed method is indeed sufficient to determine the structure of space from the inside - in theory. But viewed from the perspective of physics, it's really an epistemological result, which tells us that the structure of space is in principal knowable, even though the procedure Poincare proposed is not a practical one.

To appreciate what Poincare proposed, consider the analogy of a two-dimensional ant living on a surface (actually, it lives *in* the surface). From the outside, looking down on the ant's world, we see the creature's universe as a surface. It could be (the surface of) a sphere, a torus (the surface of a ring donut), a two-holed torus, etc. From our viewpoint, we could see which it is. But would the ant living in the surface have any way of knowing the shape? For each of those possible shapes, the ant has exactly the same freedom of movement, forwards and backwards or left and right.

Poincare asked the analogous question of our 3-D universe. We can, at least in principle, travel forwards and backwards, left and right, up and down. But as with the ant, this freedom of movement does not tell us the shape. Since we can't step outside our universe and look down on it, is there any way that we could determine the shape of our universe from the inside?

The Poincare conjecture says that the following procedure will do the trick. Imagine we set off on a long spaceship journey. (This is not how Poincare formulated the conjecture; he used the formal mathematical language of topology and geometry.) As we travel, we unreel a long string behind us. After journeying a long while, we decide it is time to head home - not necessarily the same way we went out. When we return to our starting point, we make a slipknot in our string and begin pulling the string through. One of two things can then happen. Either the noose eventually pulls down to a point or else it reaches a stage where, no matter how much we pull, it cannot shrink down any further.

Poincare's conjecture is that if we can always shrink the loop to a point, then the space we live in is the 3D analogy of the surface of a sphere; but if we can find a start/end point and a journey where the loop cannot be shrunk to a point, then our universe must have one or more "holes", much like a torus. Essentially, what is going on is that our string "snags" by going around such a hole.

Figure 1. Loop on the surface of the sphere can be tightened down to a point

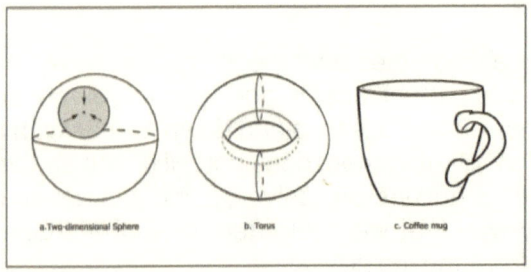

A two-dimensional sphere is simply connected be- cause any loop on the surface of the sphere can be tightened down to a point. On the other hand, a torus is not simply connected. The red loop shown here is caught around a neck of the torus and cannot be pulled any tighter.

In the nineteenth century, mathematicians came to grips with the mathematical nature of surfaces. They discovered a simple scheme for classifying surfaces using a single number, the so-called genus. Two surfaces are topologically identical, meaning it's possible to deform one into the other if and only if they have the same genus.

The simplest example of a surface is, of course, the flat, Euclidean plane. But topologists prefer to work with surfaces that are closed and bounded. The term of art is "compact" so the theory focuses instead on the sphere, which can be thought of as the plane with an extra "point at infinity." The sphere has genus zero. Roughly speaking, this means it does not have a "hole." Somewhat more precisely, it means that any closed loop drawn on the sphere can continuously shrink to a single point (Figure 1.a.)

The genus is always a non-negative integer, and there is a compact surface of each genus. A genus 1 surface has a single hole, and looks more or less like a doughnut. Such a surface is called a torus. Because it has genus 1, not every closed loop drawn on the torus can shrink to a single point, a loop that goes around the hole, for example, cannot (Figure 1.b.) In the same way the surface of a coffee mug (Figure 1.c) with a handle has genus 1. This is the source of the joke that "topologists are people who can't tell their donut from their coffee mug."

Surfaces of higher genus have correspondingly more holes, but the theory is still simple: If two compact surfaces have the same number of holes, then they are, topologically speaking, the same. The mathematical nature of compact three-dimensional spaces, or 3-manifolds as they're called, is much more complicated. There is no simple number akin to the genus that distinguishes one 3-manifold from another. Researchers have instead developed a panoply of techniques for studying 3-manifolds.

Poincaré developed an algebraic theory in which closed curves in a 3-manifold are associated with elements in a mathematical group, called the manifold's *fundamental group* in 1900. It was a crucial insight for the emergence of topology as a viable branch of mathematics. Manifolds are floppy, hard to draw and hard to visualize. Algebra, on the other hand, is concise, precise, and lends itself to symbolic manipulation. If a manifold could be uniquely described by a group, the theory of manifolds would become vastly simpler.

Alas, such a neat correspondence was not to be. Whilst it's true that if two manifolds have different fundamental groups, the manifolds are topologically different, the converse is not true: Two manifolds can have the same fundamental group yet still be different. An important exception to this was presented in 1904 by Poincaré. If the fundamental group of a 3-manifold is trivial, which happens when every closed curve can shrink to a point, then that manifold is identical to the simplest possible 3-manifold, known as the 3-sphere.

The 3-sphere is a three-dimensional analogue of the ordinary two-dimensional sphere (sometimes called the 2-sphere.) Just as the 2-sphere can be thought of as the Euclidean plane with an extra point at infinity, the 3-sphere can be thought of as Euclidean space with an extra point at infinity. It can also be viewed as the solution set in 4-dimensional space of the equation $x^2 + y^2 + z^2 + w^2 = 1$, just as the 2-sphere is the solutions set in 3-space of the equation $x^2 + y^2 + z^2 = 1$ (and the "1-sphere," or circle, is the solution set in the plane of the equation $x^2 + y^2 = 1$. In general, the *n*-sphere is the solution set of the equation $x^2_1 + x^2 + \cdots + x^2_{n+1} = 1$)

I should also point out that Poincaré didn't actually phrase the conjecture as a conjecture. He posed it as a question: Is it possible for a 3-manifold to have trivial fundamental group without being identical with the 3-sphere or being homeomorphic. The difference between a conjecture and an open question is important. To pose a statement as a conjecture, one should have strong but not necessarily convincing evidence that it is true. Poincaré's caution may have been prompted by his own bitter experience.

Four years earlier he had stated, as a theorem, a stronger claim that turned out to be false; his 1904 paper retracted that claim and gave a counterexample. In his earlier work, Poincaré was studying a different and somewhat less powerful algebraic structure associated with manifolds, known as a homology group. He had falsely claimed that an n-manifold for any dimension n is homeomorphic to the n- sphere if it has the same homology groups as the n-sphere. His 1904 counterexample was a 3-manifold with the same homology groups as the 3-sphere but with a nontrivial fundamental group.

Poincaré was the first of many people to be fooled by the elusive 3-sphere. There were numerous attempts in the twentieth century to prove the Poincaré conjecture, and several claims to have done so. Like Fermat's Last Theorem, it came to occupy mathematicians' short list of notorious problems, seemingly simple problems that nevertheless mocked the attempts of even experienced mathematicians to solve them.

As mathematicians grew more interested in higher-dimensional manifolds (which cannot be visualized in our three- dimensional universe, but which are nevertheless every bit as "real" to a topologist), they naturally wondered whether the Poincaré conjecture could be extended to these manifolds as well. If an n-dimensional, compact manifold had a trivial fundamental group, was it necessarily for an n-sphere?

At first blush, it might seem that the n-dimensional version of the Poincaré Conjecture must be much harder than the 3-dimensional version. After all, we can't even see what an n-dimensional space looks like. But the first major breakthrough on the Poincaré Conjecture came in 1960, when Stephen Smale of the Institute for Advanced Study and John Stallings at Oxford University independently proved that it was true for manifolds of 5 or more dimensions. Two decades later, in 1982, Michael Freedman proved the conjecture for $n = 4$. As a result of the deep theorems of Smale, Stallings, and Freedman, mathematicians now knew how to identify the n-sphere in every case *except* the one Poincaré had originally asked about: the case $n = 3$.

Unfortunately, there seemed to be no chance that the proofs that worked in higher dimensions could somehow be adapted to three dimensions. The extra degrees of freedom available in higher dimensions sometimes permit techniques that don't work in lower dimensions. For example, in 4 and higher dimensions, every curve can be unknotted. But as we all know from experience, knots do exist in 3-space. Thus, for example, any proof of the Poincaré Conjecture that involved untying knots would not be valid in 3 dimensions. An obstacle very much like this (only involving two-dimensional "knots") is a principal reason why Smale's and Stallings' proofs could not work in dimensions lower than 5.

This problem actually seemed to lay dormant until J. H. C. Whitehead revived interest in the conjecture, when in the 1930s he first claimed a proof and then retracted it. In the process, he discovered some interesting examples of simply-connected (indeed contractible, i.e. homotopically equivalent to a point) non-compact 3-manifolds not homeomorphic to $\mathbf{R}^3$, the prototype of which is now called the Whitehead manifold.

In the 1950s and 1960s, other mathematicians attempted proofs of the conjecture only to discover that they contained flaws. Influential mathematicians such as G. deRham, Bing, Haken, Moise, and Papakyriakopoulos attempted to prove the conjecture. In 1958 Bing proved a weak version of the Poincaré conjecture: if every simple closed curve of a compact 3-manifold is contained in a 3-ball, then the manifold is homeomorphic to the 3-sphere. Bing also described some of the pitfalls in trying to prove the Poincaré conjecture.

Włodzimierz Jakobsche showed in 1978 that, if the Bing–Borsuk conjecture is true in dimension 3, then the Poincaré conjecture must also be true. Over time, the conjecture gained the reputation of being particularly tricky to tackle. John Milnor commented that sometimes the errors in false proofs can be "rather subtle and difficult to detect." Work on the conjecture improved understanding of 3-manifolds. Experts in the field were often reluctant to announce proofs, and tended to view any such announcement with scepticism.

Hamilton's Ricci flow program, a set of equations exhibiting many similarities with the heat equation that goes back all the way to Fourier's work in 1822. This work gives manifolds a more uniform geometry and smooths out irregularities. Hamilton's technique makes use of the fact that for any kind of geometric space there is a formula called the *metric*, which determines the distance between any pair of nearby points. The Ricci flow has proven to be a very useful tool in understanding the topology of arbitrary Riemannian manifolds and in particular, as mentioned, it was a primary tool in Perelman's proof as we shall see.

Applied mathematically to this metric, the Ricci flow acts like heat, flowing through the space in question, smoothing and straightening all its bumps and curves to reveal its essential shape, the way a hair dryer shrink-wraps plastic. Geometric flows, as a class of important geometric partial differential equations, have been high-lighted in many fields of theoretical research and practical applications. They have been around at least since Mullin's work in 1956, which proposed the curve shortening flow to model the motion of idealized grain boundaries.

The Ricci flow is a geometric heat flow and can be [described] through an explanation of heat flow. The heat equation in one spatial dimension is given by $\partial_t u = \partial^2_x u$ where $u = u(x, t)$. Suppose u has a local max in space at x_0 at time t_0. Then, the second derivative $\partial^2_x u(x_0, t_0)$ will be negative, and thus by the heat equation, $u(x_0)$ is decreasing at time t_0. On the other hand, at a local minimum, the second spatial derivative will be positive, so the function will increase with time. Overall, these dual effects will act to flatten the function out, by letting heat flow from areas of high temperature to areas of low temperature.

In higher dimensions, the heat equation takes the form:

$$\partial_t u = \Delta u,$$

where Δ is the standard Laplacian. This will have the same effect of causing maxima to decrease with time, and minima to increase. In general, any differential equation of the form $\partial t u = \tilde{\Delta} u$, where $\tilde{\Delta}$ is a second order linear operator similar to the Laplacian, will have this feature of evening out the function. More precisely, we want $\tilde{\Delta}$ to be a negative operator, in the sense that $\langle u, \tilde{\Delta} u \rangle = \int u \tilde{\Delta} u$ is negative. These differential equations are called linear Parabolic equations, and they share a number of important analytic features, such as having existence of solutions for small positive time, and satisfying some type of maximum principle (meaning the size of the function can be controlled, given its initial state and boundary conditions.)

Now, a geometric heat flow is something very similar, but with some geometric invariant replacing heat. This geometric invariant will usually be some form of curvature or other, and thus the flow will tend to let curvature in the space flow around and become more evenly distributed. These flows are thus very useful, since if they can be shown to have solutions for all times, then they will tend towards a space which is evenly curved everywhere, which are much easier to understand and study.

The Ricci flow, in particular, is given by the equation:

$$\partial_t g_{ij} = -2R_{ij},$$

where g_{ij} is the metric, and R_{ij} is the Ricci curvature. The important thing to note is that the Ricci curvature is a collection of second partial derivatives of the metric, and the quantity $-R_{ij}$ on the RHS acts as a sort of Laplacian of the metric, and so we should expect similar features to the standard heat flow. Indeed, as Wikipedia explains in the article on Ricci flow, for metrics which are close to the flat metric, the Ricci flow is almost exactly the heat equation in two dimensions.

The problem with the Ricci flow is that it has the potential to develop singularities, which would cause the flow to fail. Perelman's real achievement was to get around these singularities by controlled "surgery." This means that he would allow the Ricci flow to flow for a time, until it looks like a singularity is about to form. Then, the manifold may be cut, and reglued in a way that avoids the singularity. The Ricci flow may then be turned back on. Eventually, by keeping track of the surgeries that were performed and the end result, the original manifold may be reconstructed and understood.

While geometric heat flows are closely related to linear parabolic PDEs, a key difference is that they are typically nonlinear equations. We say they are nonlinear parabolic PDEs, where the adjective "parabolic" refers to the fact that their linearizations are parabolic (perhaps after re-parametrization). For linear parabolic PDEs, we are able to show that they have solutions for all positive time; for example, starting with some fixed configuration of temperatures, the heat equation will describe their behaviour for all time. In contrast to the linear case, nonlinear parabolic PDEs tend to develop singularities in finite time, which makes their study much more involved. The goal is to find ways to avoid or control these singularities, which is the hardest part of studying a geometric heat flow.

The Ricci flow is a way, therefore, of deforming a Riemannian manifold under a nonlinear evolution equation in order to process or improve it. To be a Ricci flow, we ask a smooth one-parameter family g(t) of Riemannian metrics on a manifold M to satisfy the nonlinear partial differential equation:

$$\frac{\partial g}{\partial t} = -2\text{Ric}(g)$$

where Ric(g) is the Ricci curvature of the metric g(t), a section of $\text{Sym}^2 T^*M$.

There are a number of points to recall about Ricci curvature that may help here - first, when M is two-dimensional, Ric(g) is just the Gaussian curvature times the metric g. More generally, one should view −Ric(g) as some sort of Laplacian of g, as discussed above, so the Ricci flow is like a heat equation for g (albeit nonlinear, and liable to develop singularities). Generally, in the subject of this chapter, it is useful to view Ric(g) as a quantity which controls volume growth. In particular, it governs the change in the volume element along a geodesic.

The interpretation of Ricci flow as some sort of diffusion equation for a Riemannian metric is brought out when one computes the evolution equation governing the curvatures of a Ricci flow. For example, the scalar curvature R (which is the trace of the Ricci tensor) satisfies:

$$\frac{\partial R}{\partial t} = \Delta_{g(t)}R + 2|\text{Ric}|^2$$

Note that |Ric| is the norm of Ric using the norm induced by g (i.e. $|\text{Ric}|^2 = R_{ij}R^{ij}$)

Ricci flow has become a powerful tool for understanding the topology of manifolds, particularly 3-dimensional manifolds because it is in that dimension that the singularities of Ricci flow can be analysed to a sufficiently fine degree. A typical singularity occurs because a part of the manifold 'pinches a neck' (see fig 2. below)

Fig 2. Blowing up

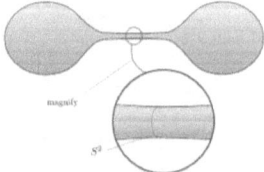

At a singularity, the curvature blows up. More precisely, the maximum sectional curvature of g(t) becomes unbounded as t increases to some singular time T. One would like to look at this flow through a microscope (i.e. rescale parabolically, thus reducing the curvature) and try to extract a limiting flow. For example, in the case of a neck pinch, the limit flow should be a cylinder $S^2 \times R$ with the S^2 part shrinking as time increases.

It turns out to be possible to find this limit only if the flow does not degenerate in a particular way. One needs to rule out 'local collapsing' where balls of a given radius, and controlled curvature, have much smaller volume than they would have in Euclidean space, or more precisely, show that:

> There exists $\kappa > 0$ such that for any $t < T$ and $r \in (0, 1]$, if the sectional curvatures of $g(t)$ are bounded in norm by r^{-2} in some geodesic ball $B(x, r)$ then the volume of $B(x, r)$ is bounded below by κr^n.

Now suppose we have extracted a limit Ricci flow (which we will also call $g(t)$), for example the shrinking cylinder mentioned above. This turns out to be defined for all $t < 0$, and when M is 3-dimensional, a result of Hamilton and Ivey tells us that it has positive curvature. It is a crucial step to analyse the geometry of limits such as $g(t)$. A further big step is to be able to relate the (scalar) curvatures at different points in space and time on such a flow and in this direction, we will be led very naturally to the following Harnack inequality which first arises in the work of Hamilton.

> For $x_1, x_2 \in M$ and times $t_1 < t_2 < 0$ on a 3-dimensional limit Ricci flow, we have:
>
> $$\frac{R(x_2, t_2)}{R(x_1, t_1)} \geq \exp\left(-\frac{d^2_{g(t_1)}(x_1, x_2)}{2(t_2 - t_1)}\right)$$

Born in Leningrad in the Soviet Union (now St. Petersburg, Russia) on June 13, 1966, Perelman was the son of an electrical engineer father who liked to challenge him with math problems, brain teasers and chess games. His mother, a math teacher, and younger sister, Elena, who also became a mathematician.

At the age of 14 Perelman was noted as the top achiever in his St. Petersburg mathematics club, and Sergey Rukshin, head of the St Petersburg Mathematical Centre for Gifted Students, began to nurture his abilities. Two years later he won a gold medal with a perfect score at the International Mathematical Olympiad in Budapest, Hungary. Perelman was something of a loner but was never perceived as hostile or unfriendly by classmates and co-workers. His interests extended beyond math to Italian opera, and he spent his small amounts of pocket money on recordings.

Perelman entered Leningrad State University at age 16 and quickly was placed in advanced geometry courses. He impressed one of his teachers, Yuri Burago, who apparently told reporters, "There are a lot of students of high ability who speak before thinking. Grisha was different. He thought deeply. His answers were always correct. He always checked very, very carefully. He was not fast. Speed means nothing. Math doesn't depend on speed. It is about *deep*." For relaxation, Perelman played table tennis and sometimes played the violin, which was also his mother's instrument.

Perelman continued straight through the programs at Leningrad State, earning the equivalent of a Ph.D. in the late 1980s after writing a dissertation on Euclidean geometry. As mentioned, he worked in the early 1990s at the Steklov Institute of Mathematics, part of the USSR Academy of Sciences. Publishing several papers on topics in geometry, he gained a reputation as a promising young scholar. In 1992 he was invited to spend a year in the United States as a guest scholar at New York University and then the State University of New York at Stony Brook. The timing was lucky, for the Russian economy was contracting rapidly in the aftermath of the fall of Communism and the breakup of the Soviet Union.

Finding the environment in America stimulating, Perelman was well liked even if some found him a little eccentric. He lived on bread (Russian black bread when he could get it), cheese, and milk, and he let his fingernails grow to a length of several inches. A hobby from back in Russia that Perelman described to friends was hunting mushrooms on hikes in the woods. A central focus of Perelman's intellectual life was a weekly lecture series at the Institute for Advanced Study at Princeton University in New Jersey, which he and Chinese colleague Gang Tian, later a key explicator of his work, attended in order to interact with the top mathematical minds in the country and the world.

At Princeton Perelman encountered mathematician William Thurston, who had developed a set of generalizations abstracted from the Poincaré conjecture and expounded upon them in lectures. Perelman also met Cornell University mathematician Richard Hamilton, and, realizing the importance of his work, approached him after one talk. "I really wanted to ask him something," he recalled to reporters. "He was smiling, and he was quite patient. He actually told me a couple of things that he published a few years later. He did not hesitate to tell me. Hamilton's openness and generosity, it really attracted me. I can't say that most mathematicians act like that."

Perelman was granted a two-year fellowship to work at the University of California at Berkeley beginning in 1993, and during one lecture he gave on campus he indicated that he had joined the hunt for a proof of Poincaré's conjecture. As suggested, in 1995, Hamilton published a paper in which he discussed a few of his ideas for completing a proof of the Poincaré. Perelman, on the other hand, thought he saw a way around obstacles hampering Hamilton's advance. In 1996, he wrote Hamilton a long letter outlining his notion, in the hope of collaborating. "He did not answer," Perelman said. "So, I decided to work alone."

As suggested, Perelman's solution was based on Hamilton's theory of Ricci flow, augmented, however, by a surgery algorithm to handle the singularities as well, using results on spaces of metrics due to Cheeger, and Gromov. Hamilton himself understood that in dimension three, surgery would be a crucial ingredient for two reasons; (a) it is the way to deal with the finite-time singularities discussed above, and (b) because of the necessity in general to do the cutting on a given three-manifold in order to find the "good metric."

In the three 2002 and 2003 papers he sketched a proof of the Poincaré conjecture and as well William Thurston's Geometrization Conjecture, a special case of which is the Poincaré conjecture. Such was the intricacy of Perelman's argument, albeit, compounded by the highly compressed manner in which he presented it, that is took over three years of effort by various groups around the world before the consensus agreement came in: he had indeed proved the Poincare Conjecture.

1. He introduced an integral functional, called the reduced L-length for paths in the space-time of a Ricci flow.

2. Using (1) he proved that for every finite time interval I, a Ricci flow of compact manifolds parameterized by I are non-collapsed where both the measure of non-collapsing and the scale depend only on the interval and the geometry of the initial manifold.

3. Using (1) and results of Hamilton's on singularity development, he classified, at least qualitatively, all models for singularity development for Ricci flows of compact three-manifolds at finite times. These models are the three-dimensional, ancient solutions (i.e., defined for $-\infty < t \leq 0$) of non-negative, bounded curvature that are non-collapsed on all scales ($r_0 = \infty$).

4. He showed that any sequence of points (x_n, t_n) in the space-time of a three dimensional Ricci flow whose times t_n are uniformly bounded above and such that the norms of the Riemannian curvature tensors $Rm(x_n, t_n)$ go to infinity (a so-called "finite-time blow-up sequence") has a subsequence converging geometrically to one of the models from (3.) This result gives neighbourhoods, called "canonical neighbourhoods", for points of sufficiently high scalar curvature. These neighbourhoods are geometrically close to corresponding neighbourhoods in non-collapsed ancient solutions as in (3).

This result gives neighbourhoods, called "canonical neighbourhoods", for points of sufficiently high scalar curvature. These neighbourhoods are geometrically close to corresponding neighbourhoods in non-collapsed ancient solutions as in (3). Their existence is a crucial ingredient necessary to establish the analytic and geometric results required to carry out and to repeat surgery.

5. From the classification in (3) and the blow-up result in (4), he showed that surgery, as envisioned by Hamilton, was always possible for Ricci flows starting with a compact three-manifold.

6. He extended all the previous results from the category of Ricci flows to a category of certain well-controlled Ricci flows with surgery.

7. He showed that, starting with any compact three-manifold, repeatedly doing surgery and restarting the Ricci flow leads to a well-controlled Ricci flow with surgery defined for all positive time.

8. He studied the geometric properties of the limits as t goes to infinity of the Ricci flows with surgery.

Let us examine these in more detail. Perelman's length function is a striking new idea, one that he has shown to be extremely powerful, for example, allowing him to prove non-collapsing results.

Let $(M, g(\tau))$, $a \leq \tau \leq T$, be a Ricci flow. The reduced L-length functional is defined on paths $\gamma : [0, \tau] \to M \times [a, T]$ that are parameterized by backwards time, namely satisfying:

$$\gamma(\tau) \in M \times \{T - \tau\} \text{ for all } \tau \in [0, \tau].$$

One considers:

$$l(\gamma) = \frac{1}{\sqrt{T}} \int_0^T \sqrt{\tau} \left(R(\gamma(\tau)) + |X_\gamma(\tau)|_g^2 \right) d\tau.$$

Here, $R(\gamma(\tau))$ is the scalar curvature of the metric $g(T - \tau)$ at the point in M that is the image under the projection into M of $\gamma(\tau)$, and $X_\gamma(\tau)$ is the projection into TM of the tangent vector to the path γ at $\gamma(\tau)$. The norm of $X_\gamma(\tau)$ is measured using $g(T - \tau)$. It is fruitful to view this functional as the analogue of the energy functional for paths in a Riemannian manifold. Indeed, there are L-geodesics, L-Jacobi fields, and a function $I_{(x, T)}(q, t)$ that measures the minimal reduced L-length of any L-geodesic from (x, T) to (q, t). Fixing (x, T), the function $I_{(x,T)}(q, t)$ is a Lipschitz function of (q, t) and is smooth almost everywhere. Perelman proves an extremely important monotonicity result along L-geodesics for a function related to this reduced length functional. Namely, suppose that W is an open subset in the time-slice $T - \tau$, and each point of W is the endpoint of a unique minimizing L-geodesic. Then he defined the reduced volume of W to be:

$$\int_W (\bar{\tau})^{-n/2} e^{-l_{(x,T)}(w, T-\bar{\tau})} dvol.$$

For each $\tau \in (0, \tau]$ let $W(\tau)$ be the result of flowing W along the unique minimal geodesics to time $T - \tau$. Then the reduced volume of $W(\tau)$ is a monotone non-increasing function of τ. This is the basis of his proof of the non-collapsing result: Fix a point (x, T) satisfying the hypothesis of non-collapsing for some $r \leq r0$. The monotonicity allows him to transfer lower bounds on reduced volume (from (x, T)) at times near the initial time (which is automatic from compactness) to lower bounds on the reduced volume times near T. From this, one proves then on-collapsing result at (x, T).

Perelman then turned to the classification of ancient three-dimensional solutions (ancient in the sense of being defined for all time $-\infty < t \leq 0$ of bounded, non-negative curvature that are non-collapsed on all scales. By geometric arguments he established that the space of based solutions of this type is compact, up to rescaling. Here the non-collapsed condition is crucial: After we rescale to make the scalar curvature one at the base point, this condition implies that the injectivity radius at the base point is bounded away from zero. There are several types of these ancient solutions.

A fixed time section is of one of the following types; (i) a compact round three-sphere or a Riemannian manifold finitely covered by a round three-sphere, (ii) a cylinder which is a product of a round two-sphere with the line or a Riemannian manifold finitely covered by this product, (iii) a compact manifold diffeomorphic to S^3 or $\mathbb{R}P^3$ that contains a long neck which is approximately cylindrical, (iv) a compact manifold of bounded diameter and volume (when they are rescaled to have scalar curvature 1 at some point), and (v) a non-compact manifold of positive curvature which is a union of a cap of bounded geometry (modulo rescaling) and a cylindrical neck (approximately S^2 times a line) at infinity.

Using the non-collapsing result and delicate geometric limit arguments, Perelman showed that in a Ricci flow on compact three-manifolds any finite-time blow-up sequence has a subsequence which, after rescaling to make the scalar curvature equal to one at the base point, converges geometrically to a non-collapsed, ancient solution of bounded, non-negative curvature. (Geometric convergence means the following: Given a finite-time blow-up sequence (x_n, t_n), after replacing the sequence by a subsequence, there is a model solution with base point and scalar curvature at the base point equal to one such that for any $R < \infty$ the balls of radius R centred at the x_n in the nth rescaled flow converge smoothly to a ball of radius R centred at the base point in the given model.) For example, we could have a component of positive Ricci curvature. According to Hamilton's result, this manifold contracts to a point at the singular time and as it does so it approaches a round (i.e., constant positive curvature) metric.

Thus, the model in this case is a round manifold with the same fundamental group. Perelman's result implies that there is a threshold so that all points of scalar curvature above the threshold have canonical neighbourhoods modelled on corresponding neighbourhoods in non-collapsed ancient solutions. This leads to geometric and analytic control in these neighbourhoods, which in turn is crucial for the later arguments showing that surgery is always possible.

Now let us turn to surgery. An n-dimensional Ricci flow with surgery is a more general object. It consists of a space-time which has a time function. The level sets of the time function are called the time-slices and they are compact n-manifolds. This space-time contains an open dense subset that is a smooth manifold of dimension $n + 1$ equipped with a vector field and a horizontal metric that make this open dense set a generalized n-dimensional Ricci flow. At the singular times, the time-slices have points not included in the open subset which is a generalized Ricci flow. This allows the topology of the time-slices to change as the time evolves past a singular time. For example, in the case just described above, when a component is shrinking to a point, the effect of surgery is to remove entirely that component.

A more delicate case is when the singularity is modelled on a manifold with a long, almost cylindrical tube in it. In this case, at the singular time, the singularities are developing inside the tube. Following Hamilton, at the singular time Perelman cuts off the tube near its ends and sews in a predetermined metric on the three-ball, using a partition of unity. The effect of surgery is to produce a new, usually topologically distinct manifold at the singular, or surgery, time.

After having produced this new closed, smooth manifold, one restarts the Ricci flow using that manifold as the initial conditions. This then is the surgery process: remove some components of positive curvature and those fibered over circles by manifolds of positive curvature (all of which are topologically standard) and surger others along two-spheres, and then restart Ricci flow.

Perelman then showed the entire Ricci flow analysis described above extends to Ricci flows with surgery, provided that the surgery is done in a sufficiently controlled manner. Thus, one is able to repeat the argument ad infinitum and construct a Ricci flow with surgery defined for all time. Furthermore, one has fairly good control on both the change in the topology and the change in the geometry as one passes the surgery times. Here then is the main result of the first two preprints taken together:

Theorem:

> Let (M, g_0) be a compact, orientable Riemannian three-manifold. Then there is a Ricci flow with surgery (M, G) defined for all positive time whose zero-time slice is (M, g_0). This Ricci flow with surgery has only finitely many surgery times in any compact interval. As one passes a surgery time, the topology of the time-slices changes in the following manner. One does a finite number of surgeries along dis-jointly embedded two-spheres (removing an open collar neighbourhood of the two-spheres and gluing three-balls along each of the resulting boundary two-spheres) and removes a finite number of topologically standard components (i.e., components admitting round metrics and components finitely covered by $S^2 \times S^1$).

Completion of the Proof of the Poincaré Conjecture

As we have already remarked, the Poincaré Conjecture follows as a special case of Thurston's Geometrization Conjecture. Thus, the results of the first two of Perelman's preprints together with the collapsing result stated at the end of the second preprint, give a proof of the Poincaré Conjecture. In a third preprint Perelman gave a different argument, avoiding the collapsing result, proving the Poincaré Conjecture but not the entire Geometrization Conjecture. One shows that if one begins with a homotopy three-sphere, or indeed any manifold whose fundamental group is a free product of finite groups and infinite cyclic groups, then the Ricci flow with surgery, which by the results of Perelman's first two preprints is defined for all positive time, becomes extinct after a finite time.

That is to say, for all τ sufficiently large, the manifold at time τ is empty. We conclude that the original manifold is a connected sum of spherical space-forms (quotients of the round S^3 by finite groups of isometries acting freely) and S^2-sphere bundles over S^1. It follows that if the original manifold is simply connected, then it is diffeomorphic to a connected sum of three-spheres, and hence is itself diffeomorphic to the three-sphere. This shows that the Poincaré Conjecture follows from this finite-time extinction result together with the existence for all time of a Ricci flow with surgery.

Let us sketch how the finite-time extinction result is established in. (There is a parallel approach in using areas of harmonic two-spheres of non-minimal type instead of area minimizing 2-disks.) We consider the case of a homotopy three-sphere M (though the same ideas easily generalize to cover all cases stated above). The fact that the manifold is a homotopy three-sphere implies that π3(M) is non-trivial. Perelman's argument is to consider a non-trivial element in $\xi \in \pi_3(M)$. Represent this element by a two-sphere family of homotopically trivial loops. Then for each loop take the infimum of the areas of spanning disks for the loop, maximize over the loops in family and then minimize over all representative families for ξ. The result is an invariant W (ξ). One asks what happens to this invariant under Ricci flow. The result (following similar arguments that go back to Hamilton) is that:

$$\frac{dW(\xi)}{dt} \leq -2\pi - \frac{1}{2} R_{min}(t) W(\xi).$$

Here $R_{min}(t)$ is the minimum of the scalar curvature at time τ. Since one of Hamilton's results using the maximum principle is that $R_{min}(t) \geq -6/(4t+\alpha)$ for some positive constant α, it is easy to see that any function satisfying Equation above goes negative in finite time. Perelman shows that the same equation holds in Ricci flows with surgery as long as the manifold continues to exist in the Ricci flow with surgery. On the other hand, W (ξ) is always non-negative. It follows that the homotopy three-sphere must disappear in finite time.

6: The Snowbird and the pitcher

A thirsty Snowbird comes across a pitcher which had been full of water. But when she puts her beak into the mouth of the pitcher she can't reach the water. Just about when she decides to give up, she comes up with an idea. She drops pebbles into the pitcher, after a while the water rises up just enough for her to quench her thirst.

From the innumerable things we've learnt in school maths the Pythagorean Theorem is one that remains with us more anything. As we leave school and for many years after, when we thing about Pythagorean Theorem, we immediately draw a right triangle in our minds eye. A triangle that has a ninety-degree angle. Most will remember that the hypotenuse is the side of the triangle directly across from the right angle. It is the only side of the triangle that is not a part of the right angle. And we learned the relationship between the sides of a right triangle (see fig 1.) When c stands for the hypotenuse and a and b are the sides forming the right angle.

The formula is:

$$a^2 + b^2 = c^2$$

Fig 1. Pythagorean triangle

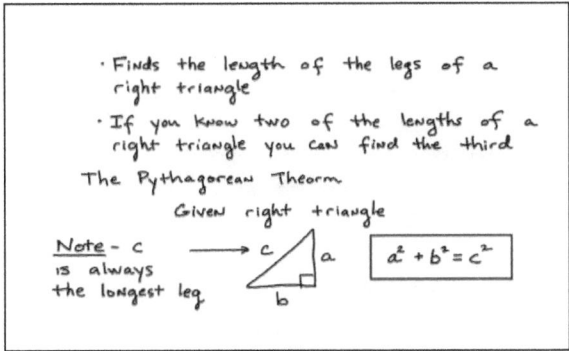

Pythagoras of Samos (c. 570 – c. 495 BC) was an Ionian Greek philosopher and the eponymous founder of the Pythagoreanism movement. His Theorem is probably the most famous of the mathematical contribution. According to legend, Pythagoras was so happy when he discovered the theorem that he offered a sacrifice of oxen. The later discovery that the square root of 2 is irrational and therefore, cannot be expressed as a ratio of two integers, greatly troubled Pythagoras and his followers.

They were devout in their belief that any two lengths were integral multiples of some unit length. Many attempts were made to suppress the knowledge that the square root of 2 is irrational. It is even said that the man who divulged the secret was drowned at sea. Regardless, this theorem was incorporated as a part of Arithmetica - a manual on number theory written in about AD 250 by Diophantus of Alexandria.

It is not known how the Greeks originally demonstrated the proof of the Pythagorean Theorem. If the methods of Book II of Euclid's *Elements* were used, it is likely that it was a dissection type of proof similar to the following:

"A large square of side a + b is divided into two smaller squares of sides a and b respectively, and two equal rectangles with sides a and b; each of these two rectangles can be split into two equal right triangles by drawing the diagonal c. The four triangles can be arranged within another square of side a + b as shown in the figures.

Fig 2. The area of the square

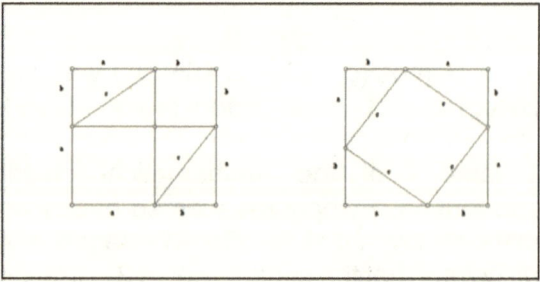

The area of the square can be shown in two different ways. As the sum of the area of the two rectangles and the squares:

$$(a+b)^2 = a^2 + b^2 + 2ab$$

As the sum of the areas of a square and the four triangles:

$$(a+b)^2 = c^2 + 4\left(\frac{ab}{2}\right) = c^2 + 2ab$$

Setting the two right hand side expressions in these equations equal, gives

$$a^2 + b^2 + 2ab = c^2 + 2ab$$

$$a^2 + b^2 = c^2$$

Therefore, the square on c is equal to the sum of the squares on a and b. (Burton 1991.) The formula that will generate all Pythagorean triples first appeared in Book X of Euclid's *Elements*:

$$x = 2mn$$

$$y = m^2 - n^2$$

$$z = m^2 + n^2$$

where n and m are positive integers of opposite parity and m > n.

Trivial solutions for Pythagoras' equation can be written such as such that x, y, and z could be any whole number, except zero. For example, $3^2 + 4^2 = 5^2$ or $5^2 + 12^2 = 13^2$ About 14-hundred years after the publication of the

Arithmetica (1637) the 17th century mathematician Pierre de Fermat considered a slightly mutated, however as we shall see, a most consequential version of the equation:

$$x^3 + y^3 = z^3$$

Apparently, Fermat created the Last Theorem while studying Arithmetica. Astonishingly, he concluded that among the infinity of numbers there were none that fitted this new form. In deed the claim was shocking. Whereas Pythagoras' equation had many possible solutions, Fermat claimed his was insoluble. Further still, believing that if the power of the equation were broadened, then these equations would also have no solutions:

$$x^3 + y^3 = z^3,$$
$$x^4 + y^4 = z^4,$$

According to Fermat, none of these equations could be solved and he noted this in the margin of his Arithmetica. To back his theorem, he had developed an argument or mathematical proof, and following the first marginal note he scribbled the most tantalising comment in the history of mathematics:

> *Cujus rei demonstrationem mirabilem sane detexi. Hanc marginis exiguitas non caperet*

or

> *I have a truly marvellous demonstration of this proposition which this margin is too narrow to contain.*

Fermat was, however, prone to bizarre leaps. "After Fermat's death, mathematicians found many similar notes," writes Simon Singh for The Telegraph. "I can provide this, but I have to feed the cat" is a memorable one.

In the early 17th century, the French polymath, Marin Mersenne who had done so much to make a careful selection of savants who met at his convent in Paris or corresponded with him from across Europe and even from as far afield as Constantinople and Transylvania threatened to cut off Fermat from the group of scholars for exactly the same reason – saying "he could but not showing he could."

The problem for us was that when Fermat said he had a proof - he either showed the proof, or the theorem was proven by someone else some many years after. But over the centuries, all of those theorems were proved, leaving just this one and a three-hundred-year history of failed attempts. Fermat believed he could prove his theorem, but he never committed his proof to paper. After his death, mathematicians across Europe tried to rediscover the proof of what became known as Fermat's Last Theorem. It was as though Fermat had buried an incredible treasure, but he had not written down the map.

Fermat, nevertheless did proved the special case $n = 4$ and with that attention turned to proving the theorem for exponents n that are primes. If the exponent "n" were not prime or 4, then it would be possible to write n either as a product of two smaller integers ($n = P*Q$) in which P is a prime number greater than 2, and then $a.n = a.P*Q = (aQ)P$ for each of a, b, and c i.e., an equivalent solution would also have to exist for the prime power P that is smaller than n, as well; or else as n would be a power of 2 greater than four and writing n=4*Q, the same argument would hold.

Fig 3. A lack of a bridge between mainland math and Fermat

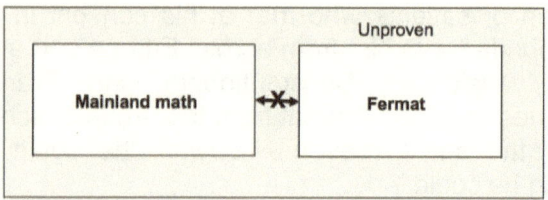

A ten-year-old boy

The history of Fermat's Last Theorem is a tale of intrigue, rivalry, rich prizes, suicide and death, involving characters who became obsessed by Fermat's accidental challenge.

Perhaps the greatest of stories starts when a 10-year old boy borrowed a book from his local library in Cambridge, England. Andrew Wiles was 10 years old in 1963, just a kid with a passion for mathematics. The young Wiles would finish his schoolwork only to make up new math problems that he solved on his own – something of a trait as we will come to see. He used the books he rented from the library as a steady stream of new math problems that would challenge him, that would teach him.

The book that had caught his eye was 'The Last Problem' by the mathematician Eric Temple Bell. The book recounted the history of Fermat's Last Theorem, the most famous problem in mathematics, which had baffled the greatest minds for over three centuries. Incidentally, the book inspired notable mathematicians including Julia Hall Robinson, the American mathematician noted for her contributions to the fields of computability theory and computational complexity theory - most notably in decision problems and John Forbes Nash best known for his work on game theory and differential geometry, and the study of partial differential equations.

Wiles took "The Last Problem" as his own and set his sights towards solving, thinking initially the problem would be easy to solve. Years later, Wiles remembers how he felt the moment he was introduced to the Last Theorem: "It looked so simple, and yet all the great mathematicians in history couldn't solve it." He said, "here was a problem that I as a 10-year-old could understand, and I knew from that moment that I would never let it go. I had to solve it." it proved anything but simple.

This led to him giving up on finding the proof and instead focus on learning mathematical techniques, with a belief he would attack the problem with greater knowledge, and with greater sophistication.

And that's precisely what he did, attending Merton College in Oxford in 1971, which is where he received his bachelor's degree majoring in mathematics and Ph.D. He was offered a string of very prestigious positions including a professor at Princeton University, a Guggenheim Fellow, Royal Society Research Professor at Oxford University.

Far from his "first love," far from the innocence of youth, in the 70's Wiles came to realize that he was far-off from the first to try reproducing Fermat's theorem. "I studied those methods," he said. "But I still wasn't getting anywhere. Then when I became a researcher, I decided that I should put the problem aside." Here then, Wiles questions his abilities. Next, he questions the prevailing techniques of mathematics. "I realized," he says "that the only techniques we had to tackle the problem had been around for 130 years.

"It didn't seem that these techniques were really getting to the root of the problem." Twenty-three years later, at the age of 33, Wiles felt he had the necessary armoury to start work on the problem once again. From 1986 to 1993, he worked on the problem in secret, in rectilinear pursuit. When asked about the level of secrecy, Wiles tells PBS, "I realized that anything to do with Fermat generates too much interest. You can't really focus yourself for years unless you have undivided concentration, which too many spectators would have destroyed."

In 2000, an older Wiles tells PBS, "I knew from that moment (as a 10-year-old) that I would never let it go, I had to solve it."

Intrigue

Wiles, like most of the mathematical community, thought Fermat was wrong. But maybe, just maybe, there is a "truly marvellous" proof out there that's much shorter than the eventual 300-page proof, we'll never know.

Fermat was born in 1601 in south-west France. Encouraged by his father to pursue a career in local government, He became a councillor at the Toulouse

Chamber of Petitions. In his spare time, he studied mathematics, a hobby which would earn him a place in history.

Known as the Prince of Amateurs, this 17th-century genius discovered the laws of probability and laid the foundations of calculus, subjects that have since revolutionised science. But Fermat's greatest contribution was to the purest and most elegant of mathematical researches - number theory, a discipline that concentrates on the properties of simple whole numbers.

For centuries, number theorists have delighted in trying to outwit each other, and Fermat was no exception. Having solved a particular problem, he would write to other mathematicians, asking if they had the ingenuity to match his solution. These challenges, and the fact that he would never reveal his own calculations, caused others a great deal of frustration. René Descartes called Fermat a "braggart", and the Englishman John Wallis referred to him as "that damned Frenchman". Fermat took particular pleasure in taunting his cousins across the Channel.

Simon Singh ('Fermat's Last Theorem, 1997) tells the story of Leonhard Euler, the greatest number theorist of the 18th century, became so frustrated that he, in 1742 asked his friend Clêrot to search Fermat's house in case some vital scrap of paper still remained. Nothing was found.

In the 19th century the most significant progress was made by Sophie Germain, a woman who, living in an era of a male world, a male chauvinism, took on the identity of a man in order to conduct her research. Although Germain's ideas eventually reached a dead-end, they generated new techniques that became valuable in tackling other problems. This was often the way with the Last Theorem - failed attempts to prove it spawned new areas of mathematical research.

One of the most intriguing stories, prior to the twentieth century concerned the most famous prize offered for a proof of the Last Theorem. It is said that toward the end of the nineteenth century Paul Wolfskehl, a German industrialist and amateur mathematician, was on the point of suicide. Some historians claim his depression was the result of a failed romance, others believe it was due to the onset of multiple sclerosis. He appointed a date for his suicide and intended to shoot himself through the head at the stroke of midnight. In the hours before his planned suicide Wolfskehl visited his library and began reading about the latest research on the Last Theorem.

Suddenly, he believed he could see a way of proving the theorem, and he became engrossed in exploring his new-found strategy. After hours of algebra Wolfskehl realised that his method had reached a dead-end, but the good news was that the appointed time of his suicide had passed. Despite his failure, Wolfskehl had been reminded of the beauty and elegance of number theory, and consequently he abandoned his plan to kill himself. Mathematics had renewed his desire for life. As a way of repaying a debt to the problem which saved his life, he re-wrote his will and bequeathed 100,000 Marks (worth $2 million in today's money) to whoever succeeded in proving Fermat.

Soon after his death in 1906, the Wolfskehl Prize was announced, generating an enormous amount of publicity and introducing the problem to the general public. Within the first year 621 proofs were sent in, most of them from amateur problem-solvers, all of them flawed.

Fiction and popular culture

Fermat's problem isn't necessarily confined to esoteric mathematics or the chivalric romance made popular in the aristocratic circles, particularly those of French origin. Fermat has repeatedly received attention in fiction and popular culture.

A sum, proved impossible by the theorem, appears in an episode of *The Simpsons*, "Treehouse of Horror VI." In the three-dimensional world in "Homer³", the sum is written out on a blackboard, just as the dimension begins to collapse. The joke is that the twelfth root of the sum does evaluate to 1922 due to rounding errors when entered into the calculator. The left hand side is odd, while 1922^{12} is even, so the equality cannot hold. The twelfth root of the left-hand side is not 1922, but approximately 1921.99999996.

A second "counterexample" appeared in a later episode, "The Wizard of Evergreen Terrace." $3987^{12} + 4365^{12} = 4472^{12}$. These agree to 10 of 44 decimal digits, but simple divisibility rules show 3987 and 4365 are multiples of 3 so that a sum of their powers is also. The same rule reveals that 4472 is not divisible by 3, so that this "equation" cannot hold either.

In similar amusement, in the Doctor Who episode "The Eleventh Hour", the Doctor transmits a proof of Fermat by typing it in just a few seconds on a laptop, to prove his genius to a collection of world leaders discussing the latest threat to the human race.

And, in theatre, Fermat's Last Tango is a stage musical by Joanne Sydney Lessner and Joshua Rosenblum. Protagonist "Daniel Keane" is a fictionalized Andrew Wiles. The characters include Fermat, Pythagoras, Euclid, Newton and Gauss, the singing, dancing mathematicians of "the aftermath" In fiction and my favourite Arthur C. Clarke and Frederik Pohl's novel The Last Theorem tells of the rise to fame and world prominence of a young Sri Lankan mathematician who devises an elegant proof of the theorem.

What good is it?

Jumping the gun, a little, Wiles' proof to Fermat has been called the proof of the century. It has been compared to the mathematical equivalent of splitting the atom or finding the structure of DNA. It is the culmination of a struggle which involved generations of mathematicians, and which has influenced every area of mathematics. Undoubtedly the solution is a triumph of modern mathematics, but for Wiles, a tremendous personal meaning. It is the realisation of a childhood dream, the end of an obsession which dominated more than just his adult life.

When in 2000 Wiles was asked how he began to look at the proof here's what he said "In my early teens I tried to tackle the problem as I thought Fermat might have tried it. I reckoned that he wouldn't have known much more math than I knew as a teenager." Then when I reached college, "I realized that many people had thought about the problem during the 18th and 19th centuries and so I studied those methods."

He goes on "but still I wasn't getting anywhere. Then when I became a researcher, I decided that I should put the problem aside. It's not that I forgot about it - it was always there but I realized that the only techniques we had to tackle it had been around for 130 years."

All this may be so, but Wiles is also a pragmatist – he goes on to tell PBS "… the problem with working on Fermat was that you could spend years getting nowhere. It's fine to work on any problem, so long as it generates interesting mathematics along the way - even if you don't solve it at the end of the day."

> "So long as it generates interesting mathematics along the way"

Ken Ribet (key to the eventual proof of Fermat) said "the main thing about Fermat is that over the centuries it gives rise to a tremendous amount of new, and fruitful mathematics. That people really develop techniques to try to solve that equation."

As well, Wiles was gracious when he said "I had this very rare privilege of being able to pursue in my adult life what had been my childhood dream. I know it's a rare privilege, but if you can tackle something in adult life that means that much to you, then it's more rewarding than anything imaginable." Clearly, Fermat is more than just the source of recognition.

But the importance of Fermat is in itself not particularly profound. It, again, in itself doesn't lead to anything that's useful that any of us know. It's not what you would consider mainstream, important, central question in modern mathematics.

Karl Gauss, considered to be one of the three greatest mathematicians of all time. The other two being Archimedes and Sir Isaac Newton summed up the attitudes of many pre-1985 professional mathematicians when in 1816 he wrote: 'I confess that Fermat, as an isolated proposition, has very little interest for me, because I could easily lay down a multitude of such propositions, which one could neither prove nor dispose.' Some-how we got lucky and managed to save Fermat's Last Theorem from its isolation by linking it to some important branches of modern mathematics, especially the theory of modular forms.'

Were we lucky?

But, Wiles is right - a good mathematical problem is in the mathematics it generates, rather than the problem itself. Without doubt, the more obviously important part comes from the methods Wiles used in the proof itself, which involved approaches and tools new to mathematics. In particular those tools could be applied to solve other difficult problems. Wiles didn't just prove Fermat, in doing so, he unlocked new doors in the field of mathematics.

It can be argued that spending time in such an "impractical" problem has more to do with their beauty than their applications. Srinivasa Ramanujan once said an equation has no meaning for him unless it expresses a thought of god. Like we appreciate a painting or a poem, the same way these problems add beauty to the art of Maths.

Mathematical theorems barely exist in a vacuum: they are of interest because they fit into a larger theory that tells us how to solve a range of problems. Whilst Fermat didn't fit into any such framework (it was not a Diophantine equation) and for that reason, no one had any real confidence that Fermat would ever be solved.

... until 1984.

Game plan

In 1949, 4 years after the end of world war II, in a war-ravaged Japan, Goro Shimura and a young Utaka Taniyama started work on the complex mathematics of modular functions. Shimura and his fellow students had to rely on each other for inspiration given the great deal of general malaise in every sector of Japan. As a by the way, Shimura tells PBS that was when he became close to Taniyama, he found him not a very careful person as a mathematician. "He made a lot of mistakes, but he made mistakes in a good direction, and so eventually, he got right answers, and I tried to imitate him."

Not surprisingly, Shimura found out that it is very difficult to make good mistakes. And as for modular forms, According to Wiles, Martin Eichler remarked that there are five fundamental operations of arithmetic: addition, subtraction, multiplication, division, and modular forms.

Eichler and Shimura developed a method to construct elliptic curves from certain modular forms. In 1955

Shimura and Taniyama suspected a link might exist between elliptic curves and modular forms, two completely different areas of mathematics. Known at the time as the Taniyama–Shimura–Weil conjecture, and (eventually) as the modularity theorem.

Fig 4. Suspected link between elliptic curves and modular forms

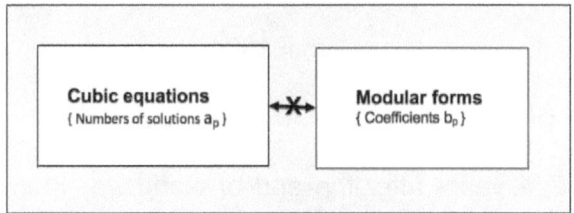

We need to hold the thought that:

> The converse notion that every elliptic curve
> has a corresponding modular form would later
> be the key to the proof of Fermat.

Wikipedia tells us the theory of Modular Forms belongs to complex analysis but the main importance of the theory has traditionally been in its connections with number theory. Also, a modular form is a complex analytic function on the upper half-plane satisfying a certain kind of functional equation with respect to the group action of the modular group, and also satisfying a growth condition.

Modular Forms are functions [singular] that provide that links between widely different fields of mathematics (see below) such as complex analysis, number theory, group theory, topology, algebra, geometry, differential equations, string theory, cryptography, and others. They were invented in the 1830s by mathematician Gauss, who was working to interpret the difficult and abstract concepts of number theory in a geometric way. Mathematical objects can be used to represent other types of mathematical objects. For instance, an algebraic equation that represents a geometric shape. A polynomial equation can be equivalent in certain ways to a topological group.

A modular form is one type of mathematical object. Essentially, it is a function over a lattice of the complex plan (or imaginary numbers.) A modular form also has some other various properties about what it does at the boundaries where it reaches infinity, and how quickly its values change.

Modular forms are versatile in that they can represent many other mathematical objects. And, when replacing a mathematical problem from one representation to another, solutions may be derived by using the machinery associated with new representation, particularly if the new representation comes with its own theorems. A variety of mathematical constructs can be represented as modular forms. In fact, there is now the "L-functions and Modular Forms Database," that lists a detailed atlas of mathematical objects that maps out the connections between them.

An application of modular forms to a domain not obviously connected to them is in the theory of lattices.

A lattice is a discrete subgroup of $\mathbb{R}^n$ that is generated by a basis of $\mathbb{R}^n$. The unimodular lattices are the ones with fundamental domains of volume 1. The even lattices are the ones where the norms of the elements are all even integers (with the usual Euclidean norm on $\mathbb{R}^n$, except in this field it's conventional to use the word *norm* to refer to the length *squared*.)

Given a lattice L, write a_k for the number of vectors in the lattice of norm k. We can form a generating function $\Theta L(q) = \sum_k a_k q^k$

Now comes the surprise: if L is an even unimodular lattice, then $\Theta_L(\exp(\pi i \tau))$, as a function of τ, is a modular form of weight n/2. Spaces of modular forms of given weight are finite dimensional. We now have a vast amount of information at our disposal on the possible norms of lattice vectors.

This gives some beautiful relationships between well-known modular forms and lattices like E_8 and the 24-dimensional Leech lattice.

It's neat how the geometrical problem of finding pockets of space into which you can squeeze lattice elements can translate into discovering modular forms. The appearance of the Leech lattice hints at Monstrous Moonshine.

We should make a brief remark as to what constitutes a mathematical object. Whilst the foundations of mathematics are concrete and precise, some things, like the word object are not rigorously defined because the meaning varies from one context to the next. An "object" is an informal term, that allow mathematicians to convey a general notion. In most cases, it's something you can act on or do something to. They are the "nouns" in the language of mathematics. A good rule of thumb, a mathematical object is something that can be labelled. The point p, the line AB, the number n, the function f, the matrix A, the group G, the manifold M, ...

And that thought?

> The converse notion that every elliptic curve has a corresponding modular form would later be the key to the proof of Fermat.

Elliptic curves are geometric objects that can be described by an equation of the form $y^2 = x^3 - px - qy$, where p and q are constants see fig 5.

The curves are naturally a group where the group law is constructed geometrically. Elliptic curves have (almost) nothing to do with ellipses, so put ellipses and conic sections out of your thoughts. Elliptic curves appear in many diverse areas of mathematics, ranging from number theory to complex analysis, and from cryptography to mathematical physics.

Formally, an elliptic curve is smooth, projective, algebraic curve of genus one, on which there is a specified point O. An elliptic curve is an abelian variety – that is, it has a multiplication defined algebraically, with respect to which it is an abelian group – and O serves as the identity element. Often the curve itself, without O specified, is called an elliptic curve; the point O is often taken to be the curve's "point at infinity" in the projective plane.

Fig 5. Elliptic curves

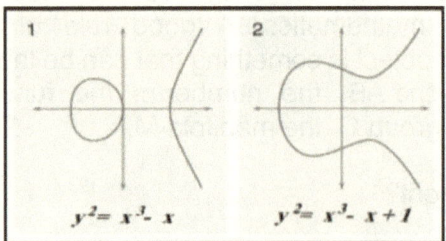

The distinguishing property of elliptic curves is that they have an "addition law". This means that given two points on an elliptic curve, there is a *natural* and *geometric* way to "add" those two points to produce a third point. This process is called "addition" because it satisfies many of the same properties that addition of integers satisfies, such as P + Q = Q + P (commutativity) and (P + Q) + R = P + (Q + R) (associativity.)

Fig 6. An example of this addition law.

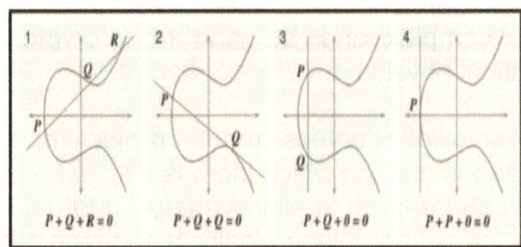

Given two distinct points P and Q on the elliptic curve, connect P and Q by a straight line. It is a fact that this straight line will intersect the curve a total of 2 times, or a total of 3 times, and these are the only possibilities. If the line does not intersect the curve in a third point, then we say that P + Q = 0. If the line does intersect the curve in a third point R, then we say that P + Q = -R.

If the line does intersect the curve in a third point R, then P + Q = -R as shown in Fig 4.1. To add a point P to itself, a line which is tangent to the curve is drawn at P. It is a fact that this straight line will intersect the curve a total of 1 time, or a total of 2 times, and these are the only possibilities. If this line does not intersect the curve in a second point, then P + P = 0 as shown in figure 6.4

If this line does intersect the curve in a second point Q, P+P=-Q as shown in figure 4.2 with the roles of P and Q swapped. Given a point P, the negative of the point, -P, is given by the unique point on the curve such that the line connecting those points does not intersect the curve in another point. In the figures above, the negative of a point is its reflection about the x-axis.

So, did work in elliptic curve and modular forms help Wiles?

In 1984, Gerhard Frey, the German mathematician from the university of Saarland in Saarbrucken, known for his work in number theory noticed that there was a connection between the equation $a^n + b^n = c^n$ and the theory of elliptic curves. More specifically, he noticed that if there was a non-trivial solution (a, b, c), then the elliptic curve $y^2 = x(x - a^n)(x+b^n)$ would have some very strange properties. Thus, a Fermat with the Taniyama – Shimura - Weil conjecture, and (eventually) as the modularity theorem.

Another thought to hold on to...

> *Frey (somehow) realized if he had a solution to Fermat's equation this would give him an elliptic curve which apparently could never be modular.*

> Albeit, extremely difficult to prove.

In 1985, Jean-Pierre Serre in Paris attempted to distil Frey's ideas, and came out with a very clear letter basically saying if you can prove this "tiny little result" which he called epsilon because the number (because in Calculus) epsilon is a small number then you can really show that on the modularity conjecture, which he called "Weil" at the time implies Fermat. So, epsilon became then became a target that Ken Ribet chassed.

Fig 7. Modularity conjecture if epsilon

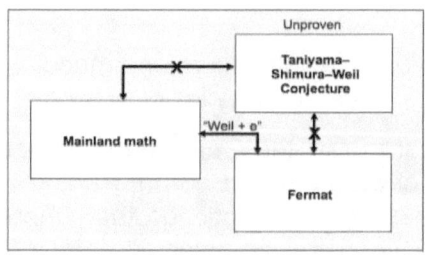

Around 1985 and 1986 Ken Ribet submitted his final grades and as he recalled, "started thinking about this problem [epsilon] first while he was on the east coast of the

US for a while, going to Europe and finding himself at an empty desk at the Max Plank Institute in Bonn." Ribet says, "to his astonishment he was able to show the simplest case for epsilon." As he tells the story, he used the very first Apple Macintosh he had access to and write to Barry Mazur at Harvard explaining what he had done. Ribet spent the following weeks to develop a more general form of epsilon, albeit with little success.

As fortune would have it, the 1986 international congress of mathematicians was held at UC Berkeley. Barry Mazur, among all well-known mathematicians were at Berkeley. Ribet, apparently approached Mazur, "Barry, I sent you this letter explaining how I did a special case and I really wonder, you know, what I have to do now in the general case?" Mazur, looks at Ribet completely quizzically, "…well, you've already done the general case."

Fig 8. The Ribet bridge between mainland math and Fermat

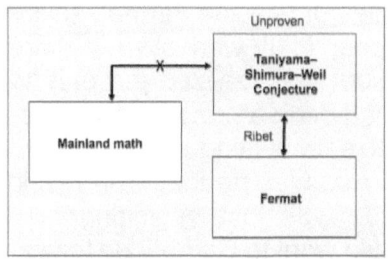

Mazur had assumed that at the time of receiving Ribet's letter, Ribet would see that with some "book keeping" Ribet would have solved the general case. Ribet remembers that they sat down at the Cafe Strada, which was at the corner of the Berkeley campus, drinking cappuccino, with the realization that in principle at least, there was no obstacle to proving this epsilon conjecture. He was well aware that he hadn't written the conjecture in any detail given he wanted to publish.

In 1986 and 1987 he lectured at the Mathematical Sciences research Institute in Berkeley, basically presenting and revising his proof. The paper was published in 1990 (later to be known as Ribet's theorem.) It should be noted that there was nothing small about Jean-Pierre Serre's "epsilon." In fact, Ribet's proof published in 1990, called *"On modular representations of $Gal(\bar{Q}/Q)$ arising from modular forms"* was some 150 pages long.

Ribet proved the modularity conjecture implied Fermat

The attack

In the 1970's Wiles puts aside Fermat, saying that "It's not that I forgot about it; it was always there." I always remembered it, but "...I realized the only techniques we had to tackle it had been around for 130 years, and it didn't seem they were really getting to the root of the problem." At this time, Fermat was no longer in fashion and at the same time Wiles was just beginning his career as a mathematician. He went to Cambridge to work on Iwasawa theory and elliptic curves with his doctoral adviser, John Coates. But at the end of the summer of 1986 Wiles remembers "... sipping iced tea at the house of a friend - this friend announces casually that Ken Ribet had proved a link between Taniyama-Shimura and Fermat."

Wiles, "*I was electrified*"

"I knew that moment that the course of my life was changing because this meant that to prove Fermat all I had to do was to prove the Taniyama-Shimura conjecture."

So, from 1986 to 1993 Wiles worked in complete isolation. He told nobody that he was embarking on a proof of Fermat. He tells PBS, "I realized that anything to do with Fermat generates too much interest. You can't really focus yourself for years unless you have undivided concentration, which too many spectators would have destroyed."

Perhaps to say the obvious, how mathematics moves forwards, slowly, and in areas that are not immediately apparent of their final utility, Wiles, tells us that at the end of the summer of '91, he was at a conference, when John Coates told him about a wonderful new paper of Matthias Flach, a student of Coates, in which he had tackled the class number formula, in fact, exactly the class number formula Wiles needed. So, Flach, using ideas Kolyvagin, had made a very significant first step in actually producing the class number formula. So, at that point, Wiles thought, 'This is just what I need. This is tailor-made for the problem.' He recalls putting aside completely the old approach and devoted himself day and night to extending Flach's result.

And again, in late spring of 1993 - Wiles tells PBS that there was still a problem. How he was in a very awkward position that he thought he had most of the curves being modular, so that was nearly enough to be content to have Fermat but there were these few families of elliptic curves that had escaped the net.

He recalls... he was sitting at his desk in May of '93, still wondering about this problem, and he was casually glancing at a paper of Barry Mazur's, and there was just one sentence which made a reference to actually what's a 19th century construction, and "... I just instantly realized that there was a trick that I could use, that I could switch from the families of elliptic curves I'd been using." He had been studying them using the prime three. He realized that he could switch and study them using the prime five. They looked to him more complicated, but he could switch from these awkward curves that he couldn't prove were modular to a different set of curves, which he had already proven were modular, and use that information to just go that one last step.

it was one morning in late May. My wife, Nada, was out with the children and I was sitting at my desk thinking about the last stage of the proof. I was casually looking at a research paper and there was one sentence that just caught my attention. It mentioned a 19th-century construction, and I suddenly realized that I should be able to use that to complete the proof. I went on into the afternoon and I forgot to go down for lunch, and by about three or four o'clock, I was really convinced that this would solve the last remaining problem. "It got to about tea time and I went downstairs and Nada was very surprised that I'd arrived so late. Then I told her I'd solved Fermat."

Shortly after learning of the proof of the epsilon conjecture, it was clear that a proof that all rational semi-stable elliptic curves are modular would also constitute a proof of

Fermat's Last Theorem. Wiles decided to conduct his research exclusively towards finding a proof for the Taniyama-Shimura conjecture. Many mathematicians thought the Taniyama-Shimura conjecture was inaccessible to proof because the modular forms and elliptic curves seem to be unrelated.

Wiles opted to attempt to "count" and match elliptic curves to counted modular forms. He found that this direct approach was not working, so he transformed the problem by instead matching the Galois representations of the elliptic curves to modular forms. Wiles denotes this matching (or mapping) that, more specifically, is a ring homomorphism:

$$R_n \to T_n$$

R is a deformation ring and *T* is a Hecke ring.

Wiles had the insight that in many cases this ring homomorphism could be a ring isomorphism. Wiles had the insight that the map between R and T is an isomorphism if and only if two abelian groups occurring in the theory are finite and have the same cardinality. This is sometimes referred to as the "numerical criterion". Given this result, one can see that Fermat's Last Theorem is reduced to a statement saying that two groups have the same order. Much of the text of the proof leads into topics and theorems related to ring theory and commutation theory.

The Goal is to verify that the map $R \to T$ is an isomorphism and ultimately that $R = T$. This is the long and difficult step. In treating deformations, Wiles defines four cases, with the flat deformation case requiring more effort to prove and is treated in a separate article in the same volume entitled "Ring-theoretic properties of certain Hecke algebra".

Gerd Faltings, in his bulletin, on p. 745. gives this commutative diagram:

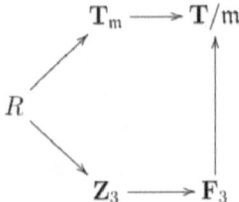

Or ultimately that $R = T$, indicating a complete intersection. Since Wiles cannot show that R=T directly, he does so through *Z3, F3* and *T/m* via lifts.

In order to perform this matching, Wiles had to create a class number formula (CNF). He first attempted to use horizontal Iwasawa theory but that part of his work had an unresolved issue such that he could not create a CNF. At the end of the summer of 1991, he learned about a paper by Matthias Flach, using ideas of Victor Kolyvagin to create a CNF, and so Wiles set his Iwasawa work aside.

Wiles extended Flach's work in order to create a CNF. By the spring of 1993, his work covered all but a few families of elliptic curves. In early 1993, Wiles reviewed his argument beforehand with a Princeton colleague, Nick Katz. His proof involved the Kolyvagin-Flach method, which he adopted after the Iwasawa method failed. In May 1993 while reading a paper by Mazur, Wiles had the insight that the 3/5 switch would resolve the final issues and would then cover all elliptic curves (this 3/5 switch). Over the course of three lectures delivered at Isaac Newton Institute for Mathematical Sciences on June 21, 22, and 23 of 1993, Wiles announced his proof of the Taniyama–Shimura conjecture, and hence of Fermat's Last Theorem. There was a relatively large amount of press coverage afterwards.

Gerd Faltings provided some simplifications to the 1995 proof, primarily in switch from geometric constructions to rather simpler algebraic ones. Because Wiles had incorporated the work of so many other specialists, it had been suggested in 1994 that only a small number of people were capable of fully understanding at that time all the details of what Wiles has done. The number is likely much larger now with the 10-day conference and book organized by Cornell et al. which has done much to make the full range of required topics accessible to graduate students in number theory.

Outline of the proof of Fermat's last theorem

Suppose we have an elliptic curve. If it is rational, then by the modularity theorem it is modular. If a curve satisfies the equation $y^2 = x(x\ a^n)(x + b^n)$, also known as a Frey curve, then the curve is nonmodular. However, if there existed integers a, b, and c, and n ≥ 3, then we would be able to construct a curve of the form $y^2 = x(x\ a^n)(x + b^n)$ that was rational. The proof of the non-modularity of the Frey curve was partially done by Jean Pierre Serre, and completed by Ribet. It later became known as Ribet's theorem. The Taniyama-Shimura conjecture, the proof of which completed the proof of Fermat's last theorem, was completed by Wiles.

In June 1993, a conference in Cambridge. Ribet recalls that Wiles has sought 3 hours from the conference because "he had something to discuss." It would take him 3 hours and needed 3 one-hour slots on consecutive days. The title of the lectures, 'Modular Forms, Elliptic Curves and Galois Representations. According to Ribet (who attended the conference), Wiles, in the first lecture was trying to prove the modularity of some class elliptic curves but the class of elliptic curves at least in the first lecture had no intersection with the class that you needed for Fermat." The second lecture "…Wiles got a lot closer and it became more and more apparent that in the third lecture he was going to do enough to do Fermat."

The crowd was growing, a lot of mathematicians bought their colleagues who happened to be in Cambridge. Others' bought their family members. Ribet, "…I think there was little doubt that he would announce the proof of Fermat."

Ribet "…I sat in the front row centre and bought my camera and snap the picture of him" and he goes on "and then you know when it actually did happen there was this frenzy of activity." On the final day, at the end of his third lecture, Wiles concluded that he had proved a general case of the Taniyama conjecture.

Then, seemingly as an afterthought, "he noted that that meant that Fermat's last theorem was true. Q.E.D."

Fig 9. Wiles noted that meant that Fermat was true

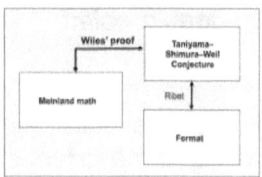

The director of the Isaac Newton Institute where the lectures were held had ordered a case of Napa Valley Brut Champagne for the end of the lecture.

Ribet remembers that a few days later he had received the manuscript, a thick 300 page single spaced typeset by TeX, the mathematics typesetting program. Ribet and others became referees each looking at different parts of the papers. Ribet was assigned a part that turned out to be problem-free while:

> Nick Katz, the Princeton mathematician found a problem. It took about 2 weeks.

It was only a few weeks before, in January of 1993, Wiles had approached Katz, a day at tea, and asked if Katz could come up to his office; "there was something he wanted to talk to me about." Katz recalls. Katz had no idea what this could be. "...I went up to his office. He closed the door. He said he thought he would be able to prove Taniyama-Shimura. I was just amazed. This was fantastic."

The flaw

In August 1993, it was discovered that the proof contained a flaw in one area. When Wiles heard about the gap, he expected that he would be able to repair the problem. However, as time went on, Wiles came to the realization that this was indeed a serious problem.

Wiles tried and failed for over a year to repair his proof. According to Wiles, the crucial idea for circumventing, rather than closing this area, came to him on 19 September 1994, when he was on the verge of giving up. Together with his former student Richard Taylor, he published a second paper which circumvented the problem and thus completed the proof. Both papers were published in May 1995 in a dedicated issue of the *Annals of Mathematics*.

Ribet, "they [Wiles and Taylor] had some real insight that permeated through the whole subject in the following 20 years." One way to say what they did is to say "modularity is contagious." In the sense that if you have something where a small piece of it is modular you can parlay that into modularity in the full.

Fig 10. The repair

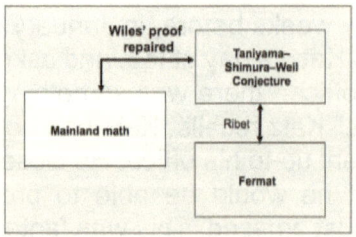

Indeed, in 1994 Andrew Wiles proved the needed special case of the Taniyama–Shimura conjecture, of which Fermat was an immediate corollary.

The Wiles-Tylor paper provides a long Bibliography and Wiles mentions the names of many mathematicians in the text. The list of some of the many other mathematicians whose work the proof incorporates includes Felix Klein, Robert Fricke, Adolf Hurwitz, Erich Hecke, Barry Mazur, Dirichlet, Richard Dedekind, Robert Langlands, Jerrold B. Tunnell, Jun-Ichi Igusa, Martin Eichler, André Bloch, Tosio Kato, Ernst S. Selmer, John Tate, P. Georges Poitou, Henri Carayol, Emil Artin, Jean-Marc Fontaine, Karl Rubin, Pierre Deligne, Vladimir Drinfel'd and Haruzo Hida and to those mathematicians who have searched (or continue to search) for a more elementary proof.

The end

www.ingramcontent.com/pod-product-compliance
Lightning Source LLC
Chambersburg PA
CBHW021817170526
45157CB00007B/2615